CONSTRUCTIVE ORDER TYPES

STUDIES IN LOGIC

AND

THE FOUNDATIONS OF MATHEMATICS

Editors

A. HEYTING, *Amsterdam*
A. MOSTOWSKI, *Warszawa*
A. ROBINSON, *New Haven*
P. SUPPES, *Stanford*

Advisory Editorial Board

Y. BAR-HILLEL, *Jerusalem*
K. L. DE BOUVÈRE, *Santa Clara*
H. HERMES, *Freiburg i/Breisgau*
J. HINTIKKA, *Helsinki*
J. C. SHEPHERDSON, *Bristol*
E. P. SPECKER, *Zürich*

NORTH-HOLLAND PUBLISHING COMPANY
AMSTERDAM · LONDON

CONSTRUCTIVE ORDER TYPES

JOHN N. CROSSLEY

Professor of Mathematics
Monash University, Melbourne, Australia

1969

NORTH-HOLLAND PUBLISHING COMPANY
AMSTERDAM · LONDON

Library of Congress Catalog Card Number 75–78330

PUBLISHERS:

NORTH-HOLLAND PUBLISHING COMPANY – AMSTERDAM
NORTH-HOLLAND PUBLISHING COMPANY LTD – LONDON

PRINTED IN THE NETHERLANDS

PREFACE

At the suggestion of J. C. Shepherdson the author first attempted to construct an analogue for order types of DEKKER and MYHILL's monograph (1960) where they studied a recursive analogue of the theory of cardinal numbers. (We give a brief survey of their work, insofar as we parallel it, in chapter 13.) Their work to some extent corresponds to TARSKI (1949) and the reader will find that the present work is related in a similar way to TARSKI (1956). This latter monograph and Dekker and Myhill's were the chief influences in the development of the present theory; for constructive order types satisfy all the finitary postulates for an ordinal algebra. (Incidentally, we note that if everywhere below we replace "recursive" by "arithmetic" then the results are even closer to those of TARSKI (1956).)

However, the problem of constructive order types was also suggested to the author by G. Kreisel and his interest arose because it was felt that the theory might throw light on the question of what is a "natural" well-ordering. We give an answer to this[01*] for ordinals less than ε_ω, the ωth solution of $\omega^\alpha=\alpha$. But although it is possible to extend the methods used here to larger ordinals (*cf.* SCHÜTTE, 1965; CROSSLEY and PARIKH, 1963; ACZEL, 1966; FEFERMAN, 1968) nevertheless we have no general method or answer for arbitrarily large recursive ordinals and this despite the fact that the first significant attack on the problem was essentially made by VEBLEN (1908). (We note that it is certainly not true that the well-orderings given by Kleene's *0* (see KLEENE, 1955) are "natural" for any

* See the notes on p. 217.

ordinal $\geq\omega^2$ under any generally accepted necessary conditions of naturalness (see, for example, ACZEL, 1966a; CROSSLEY and SCHÜTTE, 1966). Moreover the present author has felt for some time that constructing actual well-orderings is an ultimately fruitless way of achieving a definition of "natural" or of showing that there is a bound strictly less than ω_1, the first non-recursive ordinal, to all ordinals which have a natural well-ordering of their type. The reasons for this view may be expressed briefly as follows: if one can construct well-orderings for ordinals $<\alpha$ in a *uniform* way then one can construct a well-ordering of type α which is still natural (because of the uniformity). Hence one can progress further. Moreover, if the well-ordering of type α has been constructed by an explicit recursive process then $\alpha+\omega_1=\omega_1$ so α is "very small" compared with ω_1.

Because of this we do not pursue the construction of more and more functions though when the reader has read our work he may find it an interesting exercise to carry out the proofs for the results for constructive order types in our abstract with Parikh (CROSSLEY and PARIKH, 1963). Instead we study the theory from the point of view of recursion theory. This is particularly true of part two. But one fact emerges which is really quite surprising. Clearly the property of being a well-ordering is not constructive in any standard sense of "constructive", for example, the very property requires a function quantifier for its expression. Nevertheless, under the assumption of linear ordering alone one constructs *well-orderings* (when starting from finite orderings, which are, we must admit, automatically well-orderings). So we feel that it would be interesting to know what are the general properties of the functions we employ. A start has been made on this by ACZEL (in his thesis 1966) and there recursive function theory need not even be considered. We feel strongly that this approach should be continued and Kreisel and the author have suggested that categories (e.g. whose objects are linearly ordered sets with descending sequences of certain types) are an appropriate setting for such work.

From this discussion the author hopes that it will be clear that the present work is offered as a contribution to recursive function theory, though this was not how Kreisel conceived the original problem. Perhaps we should remark here that we chose the phrase 'constructive order type' because the phrases 'recursive order type' and 'computable order type' had already been used for something different, namely, (classical) order types which are the type of some recursive (or r.e.) linear ordering.

We do not restrict ourselves, in our methods of proof, to constructive means but employ whatever classical means are available. Nevertheless we do feel that a great deal of coherence has been achieved through the thoroughgoing use of (partial) recursive functions.

Of course, as Kreisel pointed out to the author in 1962, the theory should work for other sets of functions satisfying certain axioms. Now that we have amassed a quantity of results in the recursive case, it is clear that any countable set of functions closed under recursive operations will give the same results as part two. But we are also interested in developing analogues of Nerode and Myhill's work (see NERODE, 1961). We have recently found that this is possible but the generalizations go beyond sets and linearly ordered sets to embrace a multitude of algebraic systems and work on these is in progress in collaboration with Anil Nerode.

Amongst the many who have encouraged and aided me I should like to single out for special mention Alan Hamilton who read the book at the proof stages and suggested corrections and many improvements and Anil Nerode with whom I have had many hours of pleasant and fruitful discussion, especially during my visit to Cornell University in 1966.

All Souls College, Oxford.
November 2 1968

JOHN N. CROSSLEY

CONTENTS

APPENDICES

INTRODUCTION

Ordinal number theory may be approached in two distinct ways. First we may consider equivalence classes of well-ordered sets under one-one, order preserving, onto maps. On the other hand we can define ordinals set theoretically by $0=\emptyset$ (the empty set), successor of $\alpha=\alpha'=\alpha\cup\{\alpha\}$ and if σ is any set of ordinals then $\cup\sigma$ is an ordinal. This second approach was the first to be considered when it came to studying ordinal theory in recursion theory. Thus CHURCH and KLEENE (1936) set up systems of notations for an initial segment of the ordinals (the so-called *recursive ordinals*). However, though this approach has had many fruitful results it has not yet led us to a deeper insight into the structure of the ordinals. In utilizing the other approach SIERPINSKI (1958), for example, sometimes explicitly noted when the methods employed were effective in some sense: the present work is offered as a further contribution in this direction.

We shall often talk about "classical results" and by this we simply mean the results of ordinal theory as in BACHMANN (1955) or SIERPINSKI (1958) and since we often deal with analogies this has turned out to be a useful figure of speech. One classical method, namely definition by recursion, is not available to us since we have not been able to give a constructive version of the *sup* (or *lim*) operation. Moreover we believe that the best that can be achieved in this direction will be obtained by following the lines initiated in chapter 12. In order, therefore, to overcome this lack of definition by recursion we define all our operations by set-theoretic means as in SIERPINSKI (1958).

In chapter 1 we define recursive isomorphisms[02] of linearly ordered sets.

Recursive isomorphisms are one-one partial recursive maps which are order preserving. We believe it is necessary, in general, that the maps should be one-one on the whole of their domains and not just on the (possibly non-recursively enumerable) linear orderings. Perhaps it was because this was not realized at the time that RICE (1956) did not develop the theory of constructive order types for ordinal ω. With this definition of one-one for recursive isomorphisms we obtain an equivalence relation of recursive isomorphism and the equivalence classes we get are what we call *constructive order types* (C.O.T.s). We also insist, though we did not do so in our earlier work (CROSSLEY, 1965, 1966; ACZEL and CROSSLEY, 1966), that the recursive isomorphisms be order preserving wherever they are defined. The difference this makes is generally small with the exception of the results due to M. Morley in section 2.4. Since there are only countably many recursive functions and uncountably many one-one functions we see that C.O.T.s give a finer classification than (classical) order types. In section 1.2 we define an ordering R of almost the same type as the rational numbers (namely type $1+\eta$, where η is the order type of the rationals) which is recursive. Then we give (what is essentially) Cantor's proof that any two such orderings are recursively isomorphic. From this point to the end of the monograph we restrict our attention to C.O.T.s of linear orderings which can be embedded in our standard ordering R by a recursive isomorphism. This is a restriction we did not impose in CROSSLEY (1965, 1966) and ACZEL and CROSSLEY (1966) but we have found it has two advantages. First of all it removes certain features of the theory which the author considers pathological (see example IV.5.1 of CROSSLEY, 1965) and secondly it facilitates the development of a theory of losols paralleling DEKKER and MYHILL's (1960) theory of isols.[03]

Addition of C.O.T.s is defined in chapter 2 in a straightforward imitation of the classical set-theoretic definition. However we now have to ensure not merely disjointness but recursive separability. We include a useful criterion for this due to Dekker and Myhill. Once this is done it is easy to show that not only is the definition of $A+B$ satisfactory but even that there is a partial recursive functional which acts appropriately on recursive isomorphisms between representatives. Addition is associative, but not, of course, commutative as it is not commutative in the classical case. In the penultimate section of chapter 2 we establish two basic results. The first is the separation lemma, namely if $A=B+C$ and A is a linear ordering in the C.O.T. A, then we can choose $\mathsf{B}\in B$ and

$\mathsf{C} \in C$ such that the sum of B and C is actually *equal* to A and not merely recursively isomorphic to A. This lemma easily yields our directed refinement theorem (2.3.2). This theorem is used repeatedly in the sequel. The final section of chapter 2 contains M. Morley's proof of the analogue of the classical theorem of Lindenbaum and Tarski that if an order type α is similar to an initial segment of the order type β and β is similar to a final segment of α then $\alpha = \beta$. A fairly easy consequence of this is the result that $A \cdot n = B \cdot n$ if, and only if $A = B$ (where $n > 0$).

Quords, which are C.O.T.s of quasi-well-orderings, that is, linear orderings with no *recursive* descending chains, are introduced in chapter 3. We show that the condition that there be no recursive descending chains is equivalent to every non-empty recursive (or, equivalently, recursively enumerable[04]) subset having a least element. It is easy to show using SPECTOR (1955) that there are (even recursive) quasi-well-orderings which are not well-orderings though this also follows directly from Parikh's counter-example below in chapter 7.

Quords are closed under addition and, conversely, if $A = B + C$ is a quord then so too are B and C. But quords have other nice properties. For example, we can cancel on the left. But they do not have perhaps as many properties analogous to those of ordinals as one might expect (in particular see chapter 7). One question which we barely touch on and to which we would like an answer is:

What order types can recursive quords have?

For it seems that if one considers linear orderings of a certain class (say, recursive ones) with no descending chains of another (not necessarily different) class then there are *well-ordered* initial segments whose length only depends essentially on the class of functions considered (*see*, e.g. GANDY (1960)). In the present situation we merely show that recursively enumerable infinite quasi-well-orderings have types of the form $\omega + \tau$ where τ is an undetermined order type.

In chapter 4 we study our orderings of C.O.T.s and since these are most importantly employed with respect to *co-ordinals* we also define the latter. These are the C.O.T.s of well-orderings. Co-ordinals are closed under addition just like quords and the same sort of converse holds (see above). We define $A \subseteq B$ if there are representatives which are set-theoretically included in each other (with the ordering on the sets correct) and $A \leq B$ if there are representatives such that one is an initial segment of the other. $\leq^*$ is defined similarly with "final" instead of "initial". Classically, with $\subseteq, \leq$ interpreted in the obvious way, if α, β are ordinals, then $\alpha \subseteq \beta$ implies $\alpha \leq \beta$ but the proof of this fact requires non-recursive methods and the corresponding implication does not hold for co-ordinals. $\leq$ is the more important ordering and we show that it is a partial ordering of quords and a partial well-ordering of co-ordinals; but it also has the extremely useful *tree property* namely, if $A, B \leq C$ then A, B are comparable i.e. $A \leq B$ or $B \leq A$.

Finite sets are recursive and this fact is exploited in chapter 5 where we continue our study of co-ordinals. In particular, co-ordinals have unique successors, and if their classical ordinals are successor numbers then the co-ordinals have unique predecessors too. However, when we deal with limit numbers we get violent deviations from the classical theory. Thus there are uncountably many co-ordinals of type ω and these are all pairwise incomparable with respect to $\leq$. But although we have so many unreasonable co-ordinals to choose from most of our later counter-examples are based on just one and that a recursive one. We let V be the co-ordinal of a r.e. non-recursive set (of non-negative integers) ordered by magnitude. On the other hand the co-ordinal W of all the non-negative numbers ordered by magnitude is "natural" and is employed in many situations to mirror the classical results. However, we can still make use of the classical theory and a striking example is given in section 5.4 where we characterize co-ordinals which commute under addition.

Multiplication is introduced in chapter 6. We imitate the standard set-theoretic definition but code ordered pairs by single numbers in order to remain within our definition of ordering. Quords and co-ordinals are both closed under multiplication and a sort of converse (as in the case of addition) holds (with the usual trivial exceptions). Multiplication is associative and distributive over addition on one side but, as in the classical case, not commutative and not distributive on the other side.

In section 6.2 we establish a weak infinite generalization of the separation lemma and § 6.3 is devoted to a proof that $B+A=A$ if, and only if, $B \cdot W \leq A$ when A is a co-ordinal and a slightly weaker version for quords. The results of these two sections are used to obtain analogues of a number of Tarski's theorems in TARSKI (1956). However, we have not been able to obtain analogues of Tarski's theorems for C.O.T.s in general but only for quords. In particular we do not know whether the equivalence above holds for all C.O.T.s; we suspect not. Gandy has shown (May 1968) that a weaker version, namely, $2+A=A$ implies $1+A=A$ is false for arbitrary C.O.T.s. Previously HAMILTON (1968) had established the result for the definition of C.O.T.s given in CROSSLEY (1965).

Exponentiation is introduced in chapter 7, again by the classical set-theoretic definition plus coding of certain finite functions. Co-ordinals are closed under exponentiation but PARIKH (1966) showed that this is not true of quords and we include a version of his proof. The basic laws for exponents, e.g. $A^B \cdot A^C = A^{B+C}$, go through just by imitating the classical proofs. The chapter ends with a proof that if $A \geq 1$, then $B \cdot A = A$ if, and only if, B^W divides A thus mimicking the result at the end of the previous chapter though the proof here is more messy though structurally less sophisticated.

Arithmetic laws are developed in chapter 8 and here we get significant divergence from classical results. Thus half of the analogues of the monotonicity laws for ordinals break down in the cases of addition, multiplication and exponentiation. However, consideration of principal numbers for addition (or multiplication or exponentiation) i.e. co-ordinals which absorb smaller co-ordinals additively (etc.) allows us to salvage linearly (and therefore well-) ordered sets of co-ordinals which obey the analogues of the classical laws. Even without this restriction we have two techniques which yield interesting and useful results. First we can use our results of the previous chapters to obtain TARSKI (1956)-style results and secondly we use the (classical) fact that there is a unique (classical) isomorphism (onto) between two well-ordered sets of the same ordinal so that if we have a *recursive* isomorphism it must be an extension of that classical one. This gives us cancellation theorems for co-ordinals and we extend this same technique in chapter 16.

Cantor normal forms are treated in chapter 9, first in the weak sense of decomposition into sums of principal numbers for addition and then in a straight analogue of the classical case for a large class of co-ordinals. In order to effect the proof we prove the basic Euclidean algorithm, $C < A \cdot B$ implies $C = AQ + R$ where $0 \leq R < A$, by a direct argument, and likewise treat its additive and exponential analogues.

In chapter 10 we show that principal numbers for addition and multiplication are unique for ordinals $<\omega^\omega$, ω^{ω^ω}, respectively, and moreover we show that W^A, W^{W^A} are principal numbers for addition, multiplication, respectively, though in chapter 12 we shall show that not all such principal numbers are of this form. A lot of the chapter is taken up by computations which from time to time appear to require interjections of classical results to give all the information we need. It would be interesting to know if these classical interjections could be avoided. The bounds ω^ω and ω^{ω^ω} cited above seem to tie in directly with the results of EHRENFEUCHT (1957) since here we are essentially dealing with the universal formulae of his theories.

E-numbers, our analogues of the classical ε-numbers are introduced in chapter 11 and here we give the Aczel-Crossley proof that E-numbers can be characterized in an exactly analogous fashion to ε-numbers. Thus, for example, E-numbers are the solutions of $W^A = A$. As an immediate corollary of our proof we obtain the ε_ω-uniqueness for principal numbers for exponentiation.

ACZEL in his thesis (1966) introduced recursive sequence types and we introduce a related concept, constructive sequence types, in chapter 12. Because the infinite sequences we consider have uniformity properties we are able to define sums so that the co-ordinals obtained are unique. We also prove analogues of the separation lemma and the directed refinement theorem in two versions. Next in this chapter, we consider upper bounds of sets of co-ordinals. The general situation is that upper bounds exist only for countable linearly ordered sets of co-ordinals and least upper bounds exist only trivially, i.e. as maxima. On the other hand there is an abundance of what we call minimal upper bounds for any bounded linearly ordered set of co-ordinals. Finally, we prove the existence of linearly ordered sets of co-ordinals (paths) which contain one co-ordinal for each countable ordinal and are closed under addition, multiplication and exponentiation so that they are true likenesses of the classical countable ordinals.

Part two of the monograph commences with a survey of Dekker and Myhill's (and others') results on isols and recursive equivalence types. These latter are the cardinal analogues of C.O.T.s so we set out to find the analogues of isols. This we did and they rejoice in the name of *losols*. These are C.O.T.s of linearly ordered sets which are not recursively infinite (in the sense that they contain an infinite r.e. subset). In chapter 13 where we treat isols we only sketch proofs in general since either they are available elsewhere or can be reconstructed easily from their order analogues in other parts of the monograph. § 13.1 contains the basic definitions, § 13.2 exhibits several different ways of characterizing isols and § 13.3 considers the cancellation laws.

Quasi-finite C.O.T.s are introduced in chapter 14. These are C.O.T.s which are quords and whose converses are also quords (though we establish many equivalent definitions). These C.O.T.s get their name from the fact that classically finite ordered sets may be defined as those with no descending or ascending chain and in our case we insist that there should be no *recursive* ascending or descending chains. In section 2 of this chapter we show that there exist recursive quasi-finite linear orderings of type $\omega + \omega^*$. This was originally shown by Stanley Tennenbaum and the proof given here is based on a construction by C. G. Jockusch.

However, quasi-finite C.O.T.s are not closed under exponentiation though they are closed under addition and multiplication and we have not been able to establish the full "converses" of the closure results.

Losols are introduced in chapter 15. All losols are quasi-finite but the converse is false. We give the Hamilton-Nerode proof that there are $2^{\aleph_0}$ losols of type τ (where τ is any countably infinite order type) and that these may be chosen to be pairwise incomparable under $\subseteq$.

We change our definition of $\leq$ slightly for convenience. Next we show that losols are closed under addition, multiplication and exponentiation and again the pseudo-converses hold with trivial exceptions. Parikh's construction is defeated here because there being no infinite r.e. subsets implies that there are no recursive ascending or descending chains.

Chapter 16 is an extended exploitation of the fact that there is a unique order isomorphism between any two finite linearly ordered sets of the same cardinal. This culminates in a general cancellation metatheorem to the effect that $P(X)=P(Y)$ implies $X=Y$ whenever P is a non-trivial function constructed from taking converses, addition, multiplication and exponentiation using losol parameters. This brings us to our final question:

Does there exist a Myhill-Nerode-style theory (see NERODE (1961)) of extensions of functions to losols?

The answer is "Yes" and a detailed treatment of this (rather complicated) subject by Nerode and the present author is in preparation.

There are two appendices: Appendix A by P. H. G. Aczel and the author has as its main theorem:

If C is a quord and $1 \leq C$ then $CA=CB$ implies $A=B$ for *arbitrary* C.O.T.s A, B.

Appendix B contains Hamilton's proof that there are $2^{\aleph_0}$ principal numbers for multiplication not of the form W^A thus resolving a question left open in chapter 12. He also introduces the notion of an infinite product.

We have relegated the section on notation and terminology to the end on the grounds that almost all of it is standard and the section will probably only be used for reference.

PART ONE

CHAPTER 1

RECURSIVE ISOMORPHISM

1.1 By a *relation* we mean a set of ordered pairs of natural numbers, i.e. a subset of $\mathcal{N}^2$. We use upper case Gothic letters (A, B, ...) for relations. A relation A is said to be *reflexive* if

(i) $$\langle x, y\rangle \in \mathsf{A} \Rightarrow \langle x, x\rangle \in \mathsf{A} \;\&\; \langle y, y\rangle \in \mathsf{A}.$$

By the *field* of a relation A we mean

$$\{x : (\exists y)\,(\langle x, y\rangle \in \mathsf{A} \vee \langle y, x\rangle \in \mathsf{A})\}.$$

We write $C`\mathsf{A}$ for the field of a relation A. If A is a reflexive relation, then

$$C`\mathsf{A} = \{x : \langle x, x\rangle \in \mathsf{A}\}.$$

A *partial ordering* is a reflexive relation A such that

(ii) $\langle x, y\rangle \in \mathsf{A} \;\&\; \langle y, x\rangle \in \mathsf{A} \Rightarrow x = y$ (*Antisymmetry*)

and

(iii) $\langle x, y\rangle \in \mathsf{A} \;\&\; \langle y, z\rangle \in \mathsf{A} \Rightarrow \langle x, z\rangle \in \mathsf{A}$ (*Transitivity*).

A *linear ordering* is a partial ordering which also satisfies the *trichotomy law*

(iv) $$x, y \in C`\mathsf{A} \Rightarrow : x = y \vee \langle x, y\rangle \in \mathsf{A} \vee \langle y, x\rangle \in \mathsf{A}.$$

In this monograph we shall be exclusively concerned with linear orderings.

1.1.1 Definition. A relation A is said to be *isomorphic* to a relation B

if there is a one-one (possibly partial) function f such that

(i) $$C\text{‘}\mathsf{A} \subseteq \delta f \quad \text{and} \quad f(C\text{‘}\mathsf{A}) = C\text{‘}\mathsf{B}$$

and

(ii) $$\langle x, y\rangle \in \mathsf{A} \Leftrightarrow \langle f(x), f(y)\rangle \in \mathsf{B}.$$

If there is such an f we write $f{:}\,\mathsf{A} \sim \mathsf{B}$, or even simply $\mathsf{A} \sim \mathsf{B}$.

In this definition and in definitions 1.1.4 and 1.1.5 below, f is to be one-one on the whole of its domain (and not merely on the field of A). This condition ensures that in all three cases f^{-1} is well-defined on ρf, though this means that ff^{-1} may be only a restriction of the identity function in certain cases.

Clearly isomorphism is an equivalence relation. We write $\mathrm{OT}(\mathsf{A}) = \{\mathsf{B}{:}\,\mathsf{B} \sim \mathsf{A}\}$ and if $A = \mathrm{OT}(\mathsf{A})$ then A is said to be an *order type*.

1.1.2 DEFINITION. A relation A is said to be *recursive (r.e.)* if there is a recursive $\mathfrak{A}(a, b)$ ($\mathfrak{A}(a, b, c)$) such that

$$\langle a, b\rangle \in \mathsf{A} \Leftrightarrow \mathfrak{A}(a, b),$$
$$(\langle a, b\rangle \in \mathsf{A} \Leftrightarrow (\exists c)\, \mathfrak{A}(a, b, c)).$$

1.1.3 THEOREM. If A is a recursive (r.e.) relation, then $C\text{‘}\mathsf{A}$ is recursive (r.e.) provided A is reflexive.

PROOF. $C\text{‘}\mathsf{A} = \{x{:}\,\mathfrak{A}(x, x)\}$ if A is recursive
$(= \{x{:}\,(\exists z)\, \mathfrak{A}(x, x, z)\}$ if A is r.e.).

In this monograph we shall only deal with sub-relations of r.e. relations and although in confining our attention to r.e. linear orderings we shall not be treating C.O.T.s in the generality we did in our earlier papers (CROSSLEY, 1965, 1966; ACZEL and CROSSLEY, 1966) nevertheless it appears that the presentation can be made more coherent and more natural by this strategy.

1.1.4 DEFINITION. Suppose A, B are linear orderings. Then a map $p(x)$ is said to be a *recursive isomorphism* from A to B if

(o) There exist r.e. linear orderings A', B' such that $\mathsf{A} \subseteq \mathsf{A}'$ and $\mathsf{B} \subseteq \mathsf{B}'$,
(i) p is a partial recursive function,
(ii) p is one-one,

(iii) $C\text{`}A' = \delta p, p(C\text{`}A') = C\text{`}B'$ and $p(C\text{`}A) = C\text{`}B$,
(iv) $\langle x, y\rangle \in A' \Leftrightarrow \langle p(x), p(y)\rangle \in B'$.
A is *recursively isomorphic* to B if there is a map p which is a recursive isomorphism from A to B. We write $p: A \simeq B$ if p is a recursive isomorphism from A to B and $A \simeq B$ if there is a recursive isomorphism from A to B.

We claim that recursive isomorphism is an equivalence relation. The identity map is recursive, hence recursive isomorphism is reflexive. If p is a one-one partial recursive function, then p^{-1} (defined, of course, only on ρp) is also partial recursive (*see* e.g. McCarthy, 1956, p. 177). It then follows easily that if $p: A \simeq B$, then $p^{-1}: B \simeq A$. We show that recursive isomorphism is transitive. We first observe that if $p: A \simeq B$, then A is r.e. if, and only if, B is r.e. Suppose $p: A' \simeq B'$, $q: B'' \simeq C'$, $q: B \simeq C$ and $A' \supseteq A$, $B' \supseteq B$, $B'' \supseteq B$, $C' \supseteq C$ where A', B', B'', C' are r.e. Define an r.e. B^0 as follows: Let B^- be the r.e. set of ordered pairs $\langle x, y\rangle \in B' \cap B''$ with $x \neq y$ and let $B^0 = B^- \cup \{\langle x, x\rangle : x \in C\text{`}B^-\}$. Clearly B^0 is r.e.. B^0 is linear, for suppose $\langle x, y\rangle \in B^0$ and $x \neq y$, then $\langle x, y\rangle \in B' \cap B''$ and therefore $\langle y, x\rangle \notin B' \cup B'' \supseteq B' \cap B'' \supseteq B^-$. (That B^0 is a partial ordering is obvious.) Clearly $B^0 \supseteq B$ so let p' be p with range restricted to (the r.e. set) $C\text{`}B^0$ and $q'$ be $q$ with domain restricted to $C\text{`}B^0$. Then p', q' (and therefore $q'p'$) are one-one partial recursive functions and it is now trivial to check that $q'p': A \simeq C$. We can now introduce our fundamental definition.

1.1.5 Definition. If A is a linear ordering and $A = \{B : B \simeq A\}$, then A is said to be a *constructive order type* (C.O.T.) and we write $A = \mathrm{COT}(A)$.

1.1.6 Definition. A function f is said to be a *recursive permutation* if f is recursive and maps $\mathcal{N}$ one-one onto itself. We write $f: A \cong B$ if A is recursively isomorphic to B by the recursive permutation f.

We have no significant results comparable to Dekker and Myhill (1960 theorem 5) connecting the $\cong$ and $\simeq$ relations though they are easily shown to be distinct (see also corollary 1.2.5). (In fact R. I. Soare (1969) has now shown that the analogue of that theorem 5 is false for linear orderings.)

1.1.7 Theorem. (i) $A \cong B \Rightarrow A \simeq B \Rightarrow A \sim B$.
(ii) There exist linear orderings A, B such that $A \sim B$ but $A \not\simeq B$,

(iii) There exist linear orderings C, D such that $\mathsf{C} \simeq \mathsf{D}$ but $\mathsf{C} \not\cong \mathsf{D}$.

PROOF. (i) Clear from definitions 1.1.1, 4 and 6.

(ii) Let A be a r.e. non-recursive set. Then A is infinite with an infinite non-r.e. complement $\bar{\mathrm{A}}$. Let

$$\mathsf{A} = \{\langle a, a' \rangle : a, a' \in \mathrm{A} \;\&\; a \leq a'\}$$

and

$$\mathsf{B} = \{\langle b, b' \rangle : b, b' \in \bar{\mathrm{A}} \;\&\; b \leq b'\}.$$

Then $\mathsf{A} \sim \mathsf{B}$, since A, B are both well-orderings of type ω, contained in the natural (r.e.) ordering of $\mathcal{N}$. $\mathsf{A} \simeq \mathsf{B}$ implies $\bar{\mathrm{A}} = f(\mathrm{A})$ for some partial recursive f with domain including A. This implies $\bar{\mathrm{A}}$ is r.e. which contradicts the choice of A.

(iii) Let

$$\mathsf{C} = \{\langle c, c' \rangle : 0 \leq c \leq c'\}$$

and

$$\mathsf{D} = \{\langle d, d' \rangle : 1 \leq d \leq d'\}.$$

Then if $f(n) = n + 1, f : \mathsf{C} \simeq \mathsf{D}$. But $g : \mathsf{C} \cong \mathsf{D}$ implies $g^{-1}(0)$ is undefined which is in contradiction with g being a recursive permutation.

1.1.8 COROLLARY. There is a linear ordering C such that $\mathrm{OT}(\mathsf{C}) \supset \mathrm{COT}(\mathsf{C})$.

PROOF. Let C be as in the proof of the theorem and let

$$\mathsf{E} = \{\langle x, y \rangle : x \leq y \;\&\; x, y \in \bar{\mathrm{A}}\},$$

where A is as in the proof above. Then $\mathsf{E} \in \mathrm{OT}(\mathsf{C}) - \mathrm{COT}(\mathsf{C})$.

This corollary shows that C.O.T.s give a finer classification of (denumerable) relations than do (classical) order types.

1.2 An innovation (as opposed to CROSSLEY, 1965) is a uniformization of the orderings we employ. Here we shall only consider orderings which can be embedded by a recursive isomorphism in a r.e. ordering of the type of the rationals. Cantor's original paper (see CANTOR, 1915) virtually shows that every r.e. ordering is embeddable in a standard ordering R and secondly that any two standard orderings are recursively isomorphic. However we first of all show that up to recursive isomorphism it does not matter whether we consider r.e. or only recursive orderings.

1.2.1[11] THEOREM. Every r.e. linear ordering is recursively isomorphic to a recursive linear ordering (and conversely).

PROOF. Let A be a r.e. linear ordering, then there is a recursive predicate $\mathfrak{A}(x, y, z)$ such that

$$\langle x, y\rangle \in \mathsf{A} \Leftrightarrow (\exists z)\, \mathfrak{A}(x, y, z).$$

Let $A = C‘\mathsf{A}$. If A is finite there is nothing to prove since then A is finite and hence recursive. Otherwise A is infinite r.e. and there is a one-one recursive function g which enumerates A without repetitions. Since A is linear

$$(\forall x)(\forall y)(\exists z)(\mathfrak{A}(g(x), g(y), z) \vee \mathfrak{A}(g(y), g(x), z)). \tag{1}$$

Since A is antisymmetric,

$$\text{if } \mathfrak{A}(g(x), g(y), z_0) \,\&\, \mathfrak{A}(g(y), g(x), z_1) \text{ then } g(x) = g(y),$$

and hence $x = y$ and $\mathfrak{A}(g(y), g(x), z_0)$.

Now let B be the relation defined by

$$\langle x, y\rangle \in \mathsf{B} \Leftrightarrow \mathfrak{A}(g(x), g(y), \mu_z\{\mathfrak{A}(g(x), g(y), z) \vee \mathfrak{A}(g(y), g(x), z)\}).$$

Then B is recursive, by (1), and clearly $g : \mathsf{B} \simeq \mathsf{A}$.

The analogue of theorem 1.2.1 is false for arbitrary r.e. relations (see CROSSLEY, 1965, theorem I.4.5).

A linear ordering A is said to be *dense* if

$$x \neq y \,\&\, \langle x, y\rangle \in \mathsf{A} \Rightarrow (\exists z)(x \neq z \,\&\, z \neq y \,\&\, \langle x, z\rangle \in \mathsf{A} \,\&\, \langle z, y\rangle \in \mathsf{A});$$

A is said to have a first element (or minimum) a if

$$y \in C‘\mathsf{A} \Rightarrow \langle a, y\rangle \in \mathsf{A}; \quad \text{we write} \quad a = \min(\mathsf{A}).$$

First elements are clearly unique if they exist. Similarly we may define last elements.

We now introduce our standard dense linear ordering with a first element, which is 0, and no last element. We depart a little from KLEENE (1955) to get the right ordering. Given a finite non-empty sequence of natural numbers

$$a(0), \ldots, a(n-1)$$

with

$$a(n-1) \neq 0,$$

we can uniquely assign to it the *sequence number*

$$\bar{a}(n) = \prod_{i<n} p_i^{a(i)} .$$

We may also use the notation[12]

$$\bar{a}(n) = \langle a(0), \ldots, a(n-1) \rangle .$$

We denote the (number of the) empty sequence by 0. Clearly the set of all sequence numbers is recursive. We call this set Seq and we write Seq (n) if n is a sequence number. Note that, unlike Kleene, we insist that a non-empty sequence shall not end in zero.

We next define our ordering of Seq as follows: If $x, y \in$ Seq, say

$$\begin{aligned} x &= \bar{a}(m) = \langle a(0), \ldots, a(m-1) \rangle , \\ y &= \bar{b}(n) = \langle b(0), \ldots, b(n-1) \rangle , \end{aligned} \tag{2}$$

then let x^+, y^+ be the infinite sequences obtained by putting zeros after $a(m-1), b(n-1)$, respectively. Then, if $x \neq y$, we set $x^+ < y^+$ according to first differences, i.e. if

$$x^+ = \{a(i)\}_{i=0}^{\infty}, \quad y^+ = \{b(i)\}_{i=0}^{\infty}$$

and

$$a(0) = b(0), \ldots, a(k) = b(k), a(k+1) \neq b(k+1),$$

then $x^+ < y^+$ just in case $a(k+1) < b(k+1)$, otherwise $y^+ < x^+$. Note that k exists since $x \neq y$ by assumption. We set

$$\langle x, y \rangle \in \mathsf{R} \quad \text{if} \quad x^+ < y^+ \quad \text{or} \quad x = y .$$

We shall generally write

$$x \preccurlyeq y \quad \text{for} \quad \langle x, y \rangle \in \mathsf{R}$$

and

$$x \prec y \quad \text{for} \quad \langle x, y \rangle \in \mathsf{R} \,\&\, x \neq y .$$

1.2.2 THEOREM. R is a primitive recursive dense linear ordering of Seq with first element 0 and no last element.

PROOF. The only unclear part concerns the density. Suppose $x \in$ Seq then clearly $x \prec x * \langle 1 \rangle$ so R has no last element. Suppose $x \prec y$, where x, y are as in (2), and let

$$z = \langle c(0), \ldots, c(r-1) \rangle ,$$

where

$$\begin{aligned} r &= \max\{m, n\} + 1, \\ c(i) &= a(i) \quad \text{if} \quad i < m, \\ &= 0 \qquad \text{if} \quad m \le i < n, \\ &= 1 \qquad \text{if} \quad i = r - 1. \end{aligned}$$

Clearly $x \prec z \prec y$ so R is dense.

1.2.3 THEOREM. Let A be any r.e. linear ordering (of a subset of $\mathscr{N}$) then A is recursively isomorphic to a linear subordering of R. If A has a first element then we may take the recursive isomorphism to map this element onto 0.

PROOF. By theorem 1.1.3, C‘A is r.e. If A is finite then the result is trivial. Otherwise there is a (general) recursive function a which enumerates C‘A without repetitions and such that $a(0)$ is the first element of A (if there is one); similarly let r enumerate Seq. We first define a recursive function p. Let $p(0)=0$ and suppose $p(0), \ldots, p(n)$ have been defined. Then putting

$$x <_{\mathsf{A}} y \quad \text{for} \quad \langle x, y \rangle \in \mathsf{A} \,\&\, x \neq y,$$

$p(n+1)$ is chosen to be $r(m)$ where m is the least integer such that

(i) $r(m)$ is distinct from $p(0), \ldots, p(n)$,

and

(ii) either

(a) if $a(i) <_{\mathsf{A}} a(n+1) <_{\mathsf{A}} a(i+1)$ then $p(i) \prec r(m) \prec p(i+1)$ or

(b) if $a(i) <_{\mathsf{A}} a(n+1)$ for $i \le n$ then $p(i) \prec r(m)$ for $i \le n$ or

(c) if $a(n+1) <_{\mathsf{A}} a(i)$ for $i \le n$ then $r(m) \prec p(i)$ for $i \le n$.

This process is *uniform* in n and since A, R are infinite $p(n)$ is always defined, hence p is a general recursive function. Further, since $\prec$ is asymmetric, p is one-one and clearly maps $\mathscr{N}$ onto C‘A. Finally we take

$$q(x) = p(a^{-1}(x)),$$

then for some B, $q: \mathsf{A} \simeq \mathsf{B} \subseteq \mathsf{R}$ since recursive isomorphism is symmetric and transitive.

1.2.4 THEOREM. If A, B are r.e. dense linear orderings with first elements but no last elements then A is recursively isomorphic to B.

PROOF. Since A, B are dense and r.e. there exist recursive functions a, b which enumerate C‘A, C‘B, respectively, without repetitions and such that $a(0), b(0)$ are the first elements of A, B. We define the graph of a recursive isomorphism $p: \mathsf{A} \simeq \mathsf{B}$. We write a^0 for $a(0)$, b^0 for $b(0)$ and suppose

$$(a^0, b^0), \ldots, (a^n, b^n)$$

have been defined.

If n is even we let a^{n+1} be $a(m)$ where m is the least integer such that

$$a(m) \notin \{a^0, \ldots, a^n\}$$

and let b^{n+1} be $b(k)$ where k is the least integer such that

(i) $b(k) \notin \{b^0, \ldots, b^n\}$

and

(ii) either

(a) if $a^i <_{\mathsf{A}} a(m) <_{\mathsf{A}} a^{i+1}$ then $b^i <_{\mathsf{B}} b(k) <_{\mathsf{B}} b^{i+1}$ or

(b) if $a^i <_{\mathsf{A}} a(m)$ for $i \leq n$ then $b^i <_{\mathsf{B}} b(k)$ for $i \leq n$ or

(c) if $a(m) <_{\mathsf{A}} a^i$ for $i \leq n$ then $b(k) <_{\mathsf{B}} a^i$ for $i \leq n$.

(Here $<_{\mathsf{B}}$ is defined analogously to $<_{\mathsf{A}}$ above.) Since B is dense and has no last element such a k always exists.

If n is odd then we proceed as for n being even except that we read "a" for "b" and *vice versa* throughout the above. Since each $a(i)$ and $b(j)$ is chosen once and only once the map $p: a^n \to b^n$ is one-one and maps C'A onto C'B. Further, since the procedure is uniform p is partial recursive. Finally p is order preserving by construction. This completes the proof.

1.2.5 Corollary. If A, B are recursive dense linear orderings (with first but no last elements) of the whole of $\mathcal{N}$ then $\mathsf{A} \cong \mathsf{B}$.

1.2.6 Corollary. Any subordering of a r.e. linear ordering can be embedded in R by a recursive isomorphism.

We shall show in chapter 3 that there are linear orderings which cannot be embedded in R by a recursive isomorphism, however in the author's earlier work such orderings were only found useful for giving examples of pathologies (e.g. Crossley, 1965, examples IV.5.1 and VIII.1.14).

CHAPTER 2

ADDITION

2.1 Since all (suborderings of) r.e. linear orderings can be embedded in our standard ordering R by a recursive isomorphism we shall restrict our attention from now on to C.O.T.s of suborderings of R unless otherwise stated. If there is an $\mathsf{A} \in A$ such that $\mathsf{A} \subseteq \mathsf{R}$ then we write[21] $A \subseteq R$ where we set

$$R = \mathrm{COT}(\mathsf{R})$$

and

$$\mathscr{R} = \{A : A \subseteq R\}.$$

We call $\mathscr{R}$ the collection of all *standard* C.O.T.s when we wish to distinguish $\mathscr{R}$ from the collection of all C.O.T.s as defined in CROSSLEY (1965).

Consider any subset A of Seq, then A *generates* the linear ordering A defined by

$$\langle x, y\rangle \in \mathsf{A} \Leftrightarrow x, y \in \mathrm{A} \;\&\; x \preccurlyeq y\,.$$

We shall therefore adopt the convention of conflating a subset of Seq with the ordering it generates. But if there is the possibility of confusion then we shall write [A] for the ordering the set A generates.

If A and B are arbitrary orderings, then the *ordinal sum* of A and B is defined by

$$\mathsf{A} \dotplus \mathsf{B} = \mathsf{A} \cup \mathsf{B} \cup (\mathrm{C}`\mathsf{A} \times \mathrm{C}`\mathsf{B})\,.$$

If the ordinal sum of two order types is defined as the order type of the ordinal sum of arbitrary representatives of the given order types, then this definition is not, in general, unique. This is because the fields of the representative orderings may have a non-empty intersection. But if we define the order type of the sum in terms of representatives which do have disjoint fields, then the definition is unique (*cf.* RUSSELL and WHITEHEAD, 1927, pp. 341, 345 *160.48). Two relations are said to be ***strictly disjoint*** if their fields are disjoint. However, we observe that if A, B are reflexive, then

$$\mathsf{A} \cap \mathsf{B} = \emptyset \Leftrightarrow C`\mathsf{A} \cap C`\mathsf{B} = \emptyset,$$

or using the notation introduced above

$$[A] \cap [B] = \emptyset \Leftrightarrow A \cap B = \emptyset$$

when $A, B \subseteq \text{Seq}$. Now, in order to define a recursive version of ordinal sum we require "recursive disjointness" i.e. $C`\mathsf{A}$ and $C`\mathsf{B}$ must be contained in sets which are effectively disjoint. If this is not the case then the following situation arises: Let A be a r.e. non-recursive set and let B be a r.e. set containing $\bar{A}$. Then there is no (partial) recursive function defined on $\mathscr{N}$ which agrees with the functions f, g on their domains of definition where $f(x)=x$ if $x \in A$, otherwise undefined and $g(x)=x+1$ if $x \in B$, otherwise undefined. Hence there can be no (partial) recursive function defined on $C`(\mathsf{A} \dotplus \mathsf{B})$ if $C`\mathsf{A}=A$ and $C`\mathsf{B}=\bar{A}$, even though A and B are strictly disjoint.

2.1.1 DEFINITION. A is *separable* from B if there are disjoint r.e. sets A', B' such that $C`\mathsf{A} \subseteq A'$ and $C`\mathsf{B} \subseteq B'$. If A is separable from B we write

$$\mathsf{A})(\mathsf{B}$$

and we also write

$$A)(B \quad \text{for} \quad [A])([B].$$

2.1.2 THEOREM. If $\mathsf{A}, \mathsf{B} \subseteq \mathsf{R}$ and

(i) every element in $C`\mathsf{A}$ precedes (under $\prec$) every element of $C`\mathsf{B}$ and

(ii) either (a) B has a first element
or (b) A has a last element

then

$$\mathsf{A})(\mathsf{B}.$$

PROOF. We treat only (a) and leave (b) to the reader. Let b be the first element of B, then the sets

$$A' = \{x : x \prec b\} \quad \text{and} \quad B' = \{x : b \preccurlyeq x\}$$

are recursive and disjoint and clearly contain C‘A and C‘B since if any element of A precedes every element of B then it certainly precedes b.

This theorem will be used frequently in the sequel, especially when we are dealing with C.O.T.s of well-orderings. The following criterion is also useful.

2.1.3 THEOREM. A) (B if, and only if, there is a partial recursive function p such that if A=C‘A and B=C‘B then

(i) $\delta p \supseteq A \cup B$, $\rho p \subseteq \{0, 1\}$

and

(ii) if $x \in A \cup B$ then

$$x \in A \Leftrightarrow p(x) = 0 \quad \text{and} \quad x \in B \Leftrightarrow p(x) = 1.$$

PROOF. If A) (B, then by definition 2.1.1 there are r.e. sets A′, B′ such that

$$A' \supseteq A, \ B' \supseteq B \quad \text{and} \quad A' \cap B' = \emptyset.$$

For an arbitrary r.e. set S let c_S be the partial recursive function such that

$$c_S(x) = 1 \quad \text{if} \quad x \in S, \quad \text{otherwise undefined.}$$

Then $x \in A'$ imples $1 \dot{-} c_{A'}(x) = 0$ and $x \in B'$ implies $c_{B'}(x) = 1$. Let T_A be a Turing machine which calculates $1 \dot{-} c_{A'}(x)$ and let T_B be a Turing machine which calculates $c_{B'}(x)$. Further, let

$T(m, n)$ = the number (represented) on the tape of the Turing machine T at the mth step[22] in the calculation for argument n.

Now let a new machine T_0 be defined such that $T_0(m, n)$ is as follows:

(i) If T_A, T_B have not halted before the mth step for argument n, then $T_0(2m+1, n) = T_A(m+1, n)$,

(ii) If T_A has not halted before the $(m+1)$th step and T_B has not halted before the mth step, then $T_0(2m+2, n) = T_B(m+1, n)$,

(iii) If T_A halts at the mth step and T_B has not halted before the mth step, then T_0 halts at the $(2m+1)$th step,

(iv) If T_B halts at the mth step and T_A has not halted before the $(m+1)$th step, then T_0 halts at the $(2m+2)$nd step.

Let $p(x)$ be the function defined by the machine T_0. Then p is partial recursive and satisfies the conclusion of the theorem since, for an argument in $\mathrm{A} \cup \mathrm{B}$, T_{A} halts if, and only if, T_{B} does not.

Conversely, let

$$\mathrm{A}' = \{x : p(x) = 0\} \quad \text{and} \quad \mathrm{B}' = \{x : p(x) = 1\}.$$

Then A′ and B′ are r.e. and disjoint and by hypothesis $\mathrm{A} \subseteq \mathrm{A}'$ and $\mathrm{B} \subseteq \mathrm{B}'$. This completes the proof.

It is very convenient to be able to produce separable representatives of given C.O.T.s and we therefore introduce some notation to this end. We recall that if $x, y \in \mathrm{Seq}$ then

$$x = \langle (x)_0, \ldots, (x)_m \rangle \quad \text{and} \quad y = \langle (y)_0, \ldots, (y)_n \rangle,$$

where

$$m = lh(x) + 1 \quad \text{and} \quad n = lh(y) + 1 \quad \text{and}$$
$$x * y = \langle (x)_0, \ldots, (x)_m, (y)_0, \ldots, (y)_n \rangle.$$

For each natural number i we let $\hat{\imath}$ be the one-one partial recursive function defined by

$$\begin{aligned} &\hat{\imath}(x) = \langle i \rangle * x \quad \text{if} \quad x \in \mathrm{Seq}, x \neq 0, \\ &\hat{0}(0) = 0, \\ &\quad \hat{\imath}(x) \text{ is undefined, otherwise.} \end{aligned}$$

We also set

$$\hat{\imath}\mathrm{A} = \{\hat{\imath}(x) : x \in \mathrm{A}\} \quad \text{if} \quad \mathrm{A} \subseteq \mathrm{Seq},$$

and

$$\hat{\imath}\mathsf{A} = [\hat{\imath}\mathrm{A}] \quad \text{if} \quad \mathsf{A} = [\mathrm{A}] \subseteq \mathsf{R}.$$

We also require the inverses of these maps; we shall denote them by $\check{\imath}$, so

$$\begin{aligned} \check{\imath}(x) &= x/2^i \quad \text{if} \quad x \in \hat{\imath}\ \mathrm{Seq}, x \neq 0, \\ &= 0 \quad \text{if} \quad x = 0, \\ &\text{is undefined otherwise.} \end{aligned}$$

The orderings we are particularly interested in are those of the form $\hat{\imath}\mathsf{R}$ and we therefore introduce the notation

$$x \leqslant_i y \Leftrightarrow \langle x, y \rangle \in \hat{\imath}\mathsf{R}$$

so

$$x \leqslant_i y \Leftrightarrow (\exists x')(\exists y')(x = \hat{\imath}(x') \;\&\; y = \hat{\imath}(y') \;\&\; x' \leqslant y').$$

2.1.4 THEOREM. (i) $x \preccurlyeq y \Leftrightarrow \hat{\imath}(x) \preccurlyeq_i \hat{\imath}(y)$,

(ii) $A = \mathrm{COT}(\hat{\imath}\mathsf{A})$ for any $\mathsf{A} \in A$.

PROOF. Left to the reader.

2.1.5 THEOREM. If $i \neq j$, then $\hat{\imath}\,\mathsf{R})\,(\hat{\jmath}\,\mathsf{R}$.

PROOF. Let $p(x) = 0$ if $x \in \hat{\imath}$ Seq,
$= 1$ if $x \in \hat{\jmath}$ Seq,
is undefined otherwise.

Now apply theorem 2.1.3.

2.1.6 COROLLARY. Any two (standard) C.O.T.s have separable representatives.

PROOF. $A, B \in \mathscr{R}$ implies there exist $\mathsf{A} \in A$ and $\mathsf{B} \in B$ such that $\mathsf{A} \subseteq \mathsf{R}$ and $\mathsf{B} \subseteq \mathsf{R}$. By theorem 2.1.5 it follows that $\hat{0}\mathsf{A})\,(\hat{1}\mathsf{B}$ and the corollary then follows from theorem 2.1.4(ii).

2.2.

2.2.1 LEMMA. Let $\mathsf{A}_0 \simeq \mathsf{A}_1$, $\mathsf{B}_0 \simeq \mathsf{B}_1$, $\mathsf{A}_0)\,(\mathsf{B}_0$ and $\mathsf{A}_1)\,(\mathsf{B}_1$, then

$$\mathsf{A}_0 \hat{+} \mathsf{B}_0 \simeq \mathsf{A}_1 \hat{+} \mathsf{B}_1 .$$

PROOF. Let $A_i = C`\mathsf{A}_i$, $B_i = C`\mathsf{B}_i$ $(i = 0, 1)$. By hypothesis there exist recursive isomorphisms p, q and linear orderings A_i', B_i' such that

$$p : \mathsf{A}_0' \simeq \mathsf{A}_1' \quad \text{and} \quad q : \mathsf{B}_0' \simeq \mathsf{B}_1' ,$$

where $\mathsf{A}_i \subseteq \mathsf{A}_i'$, $\mathsf{B}_i \subseteq \mathsf{B}_i'$ $(i = 0, 1)$ and the A_i', B_i' are r.e. Further there are r.e. orderings A_i'', B_i'' such that

$$\mathsf{A}_i \subseteq \mathsf{A}_i'', \mathsf{B}_i \subseteq \mathsf{B}_i'' \quad \text{and} \quad \mathsf{A}_0'' \cap \mathsf{A}_1'' = \mathsf{B}_0'' \cap \mathsf{B}_1'' = \emptyset .$$

Let p' be the partial recursive function defined by

$$p'(x) = p(x) \quad \text{if} \quad x \in \delta p \cap A_0' \cap A_0'' \quad \text{and} \quad p(x) \in A_1' \cap A_1'',$$
is undefined otherwise

(where $A_i = C`\mathsf{A}_i$, etc.).

Similarly let q' be the partial recursive function defined by

$$q'(x) = q(x) \quad \text{if} \quad x \in \delta q \cap B_0' \cap B_0'' \quad \text{and} \quad q(x) \in B_1' \cap B_1'',$$
is undefined otherwise.

Let r be the partial recursive function defined by

$$\begin{aligned} r(x) &= p(x) \quad \text{if} \quad p'(x) \quad \text{is defined}, \\ &= q(x) \quad \text{if} \quad q'(x) \quad \text{is defined}, \\ &\text{is undefined otherwise}. \end{aligned}$$

Since p' and q' have disjoint domains and ranges r is well-defined. Clearly

$$r:\mathsf{A}_0^+ \dotplus \mathsf{B}_0^+ \simeq \mathsf{A}_1^+ \dotplus \mathsf{B}_1^+ \quad \text{and} \quad r:\mathsf{A}_0 \dotplus \mathsf{B}_0 \simeq \mathsf{A}_1 \dotplus \mathsf{B}_1,$$

where A_0^+, A_1^+, B_0^+, B_1^+ are the r.e. relations

$$\mathsf{A}_0 \cap (\delta p')^2,\ r(\mathsf{A}_0),\ \mathsf{B}_0 \cap (\delta q')^2,\ r(\mathsf{B}_0).$$

This completes the proof.

Now $\hat{0}\,\mathrm{Seq}\,)(\,\hat{1}\,\mathrm{Seq}$ since these two sets are recursive and disjoint hence we can define for $\mathsf{A}, \mathsf{B} \subseteq \mathsf{R}$,

$$\mathsf{A} + \mathsf{B} = \hat{0}\mathsf{A} \dotplus \hat{1}\mathsf{B}$$

and we have the following strengthening of lemma 2.2.1.

2.2.2 THEOREM. There is a partial recursive functional Φ such that if $p:\mathsf{A}_0 \simeq \mathsf{A}_1$ and $q:\mathsf{B}_0 \simeq \mathsf{B}_1$, then

$$\Phi(p, q):\mathsf{A}_0 + \mathsf{B}_0 \simeq \mathsf{A}_1 + \mathsf{B}_1.$$

PROOF. Set
$$\begin{aligned} \Phi(p, q)(x) &= \hat{0}p(\overline{0}(x)) \quad \text{if} \quad x \in \hat{0}\,\mathrm{Seq}, \\ &= \hat{1}q(\overline{1}(x)) \quad \text{if} \quad x \in \hat{1}\,\mathrm{Seq}, \\ &\text{is undefined otherwise.} \end{aligned}$$

Theorem 2.2.2 allows us to make the following definition.

2.2.3 DEFINITION. $A + B = \mathrm{COT}(\mathsf{A} + \mathsf{B})$, where $\mathsf{R} \supseteq \mathsf{A} \in A$ and $\mathsf{R} \supseteq \mathsf{B} \in B$.

We note that if $\mathsf{A}\,)(\,\mathsf{B}$ then $\mathsf{A} \dotplus \mathsf{B} \in \mathrm{COT}(\mathsf{A}) + \mathrm{COT}(\mathsf{B})$ by lemma 2.2.1. *Notation*: $0 = \mathrm{COT}(\emptyset)$.

2.2.4 THEOREM. (i) $A + 0 = 0 + A = A$,
(ii) $A + B = 0 \Leftrightarrow A = B = 0$.

PROOF. Left to the reader.

2.2.5 THEOREM. $+$ is associative, viz. for all $A, B, C \in \mathscr{R}$,

$$(A+B)+C = A+(B+C).$$

PROOF. We need to show that if $\mathsf{A}, \mathsf{B}, \mathsf{C} \subseteq \mathrm{R}$ and $\mathsf{A}, \mathsf{B}, \mathsf{C} \in A, B, C$, respectively, then

$$\hat{0}(\hat{0}\mathsf{A} \dotplus \hat{1}\mathsf{B}) \dotplus \hat{1}\mathsf{C} \simeq \hat{0}\mathsf{A} \dotplus \hat{1}(\hat{0}\mathsf{B} \dotplus \hat{1}\mathsf{C}).$$

So set

$$\begin{aligned} p(x) &= \hat{0}(y) \quad \text{if} \quad x = \hat{0}\hat{0}(y) \ \& \ \mathrm{Seq}(y), \\ &= \hat{1}\hat{0}(y) \quad \text{if} \quad x = \hat{0}\hat{1}(y) \ \& \ \mathrm{Seq}(y), \\ &= \hat{1}\hat{1}(y) \quad \text{if} \quad x = \hat{1}(y) \ \& \ \mathrm{Seq}(y), \\ &\text{is undefined otherwise.} \end{aligned}$$

We leave the reader to verify that p is the required recursive isomorphism.

We know classically that $1+\omega \neq \omega+1$, so addition of C.O.T.s is not commutative.

2.3 In the previous section we were concerned with putting C.O.T.s together by means of sums, in this section we prove a useful lemma and an important theorem showing that we can, so to speak, take the C.O.T.s apart again.

2.3.1 LEMMA. (SEPARATION LEMMA) If $A = B + C$ and $\mathsf{A} \in A$, then there exist $\mathsf{B} \in B$ and $\mathsf{C} \in C$ such that

$$\mathsf{B})(\mathsf{C} \quad \text{and} \quad \mathsf{B} \dotplus \mathsf{C} = \mathsf{A}.$$

PROOF. By definition 2.2.3, there exist $\mathsf{B}', \mathsf{C}' \subseteq \mathrm{R}$ with $\mathsf{B}' \in B$ and $\mathsf{C}' \in C$ such that

$$\mathsf{B}' + \mathsf{C}' \in A, \quad \text{say} \quad \mathsf{B}' + \mathsf{C}' = \mathsf{A}'.$$

Since $\mathsf{A}' \in A$ there is a recursive isomorphism

$$p: \mathsf{A}' \simeq \mathsf{A}.$$

Let

$$\mathsf{B} = p(\hat{0}\mathsf{B}') \quad \text{and} \quad \mathsf{C} = p(\hat{1}\mathsf{C}').$$

Since $\hat{0}\mathsf{B}' \dotplus \hat{1}\mathsf{C}' = \mathsf{A}'$ we have at once

$$\mathsf{B} \dotplus \mathsf{C} = \mathsf{A}.$$

Finally, $\mathsf{B})(\mathsf{C}$ since

$$C\text{‘}\mathsf{B} \subseteq p(\hat{0}\ \mathrm{Seq}))(p(\hat{1}\ \mathrm{Seq}) \supseteq C\text{‘}\mathsf{C}.$$

This completes the proof.

Notation. $\mathsf{A}[\mathrm{B} = \{\langle x, y\rangle : \langle x, y\rangle \in \mathsf{A} \,\&\, x, y \in \mathrm{B}\}$.

2.3.2 THEOREM. (DIRECTED REFINEMENT THEOREM) If $A+C=B+D$, then there is an E such that
either

$$A = B + E \quad \text{and} \quad E + C = D,$$

or

$$A + E = B \quad \text{and} \quad C = E + D.$$

PROOF. If $D=0$, let $E=C$. Otherwise we may assume $D \neq 0$. Let $\mathsf{A} \in A$ and $\mathsf{C} \in C$ where $\mathsf{A})(\mathsf{C}$ and $\mathsf{A} \mathbin{\hat{+}} \mathsf{C} \subseteq \mathsf{R}$. Then by the separation lemma (2.3.1) there are relations $\mathsf{B} \in B$ and $\mathsf{D} \in D$ such that

$$\mathsf{B})(\mathsf{D} \quad \text{and} \quad \mathsf{A} \mathbin{\hat{+}} \mathsf{C} = \mathsf{B} \mathbin{\hat{+}} \mathsf{D}.$$

Let $\mathrm{A} = C\text{‘}\mathsf{A}$, $\mathrm{B} = C\text{‘}\mathsf{B}$, $\mathrm{C} = C\text{‘}\mathsf{C}$ and $\mathrm{D} = C\text{‘}\mathsf{D}$. Then

$$\mathrm{A} \cup \mathrm{C} = \mathrm{B} \cup \mathrm{D}.$$

We consider two cases separately.
Case 1. If $\mathrm{D} \cap \mathrm{A} \neq \emptyset$, let

$$\mathrm{E} = \mathrm{A} - \mathrm{B} \quad \text{and} \quad \mathsf{E} = \mathsf{D}[\mathrm{E}.$$

By construction, $\mathrm{A} \subseteq \mathrm{B} \cup \mathrm{E}$ and $\mathrm{E} \subseteq \mathrm{D}$, therefore

$$\mathsf{B})(\mathsf{E} \quad \text{and} \quad \mathsf{E})(\mathsf{C}.$$

Now $\mathsf{B} \subseteq \mathsf{A} \mathbin{\hat{+}} \mathsf{C}$ and $\mathsf{E} \subseteq \mathsf{A} \mathbin{\hat{+}} \mathsf{C}$; further,

$$x \in \mathrm{B} \text{ and } y \in \mathrm{E} \text{ imply } y \in \mathrm{D} \text{ and } \langle x, y\rangle \in \mathrm{B} \times \mathrm{D} \subseteq \mathsf{A} \mathbin{\hat{+}} \mathsf{C}.$$

Thus

$$\mathsf{B} \mathbin{\hat{+}} \mathsf{E} \subseteq (\mathsf{A} \mathbin{\hat{+}} \mathsf{C})[\mathrm{A} = \mathsf{A}.$$

Conversely, if $\langle x, y\rangle \in \mathsf{A}$, then one of

$$x, y \in \mathrm{B}; \quad x, y \in \mathrm{E}; \quad x \in \mathrm{B} \,\&\, y \in \mathrm{E} \tag{1}$$

must hold since $x \in \mathsf{E} \,\&\, y \in \mathsf{B}$ contradicts

$$(\mathsf{D} \times \mathsf{B}) \cap (\mathsf{B} \dagger \mathsf{E}) = \emptyset \;\&\; \mathsf{E} \subseteq \mathsf{D}.$$

In all three cases in (1) $\langle x, y\rangle \in \mathsf{B} \dagger \mathsf{E}$. Hence $\mathsf{A} \subseteq \mathsf{B} \dagger \mathsf{E}$ and therefore

$$\mathsf{A} = \mathsf{B} \dagger \mathsf{E}.$$

Similarly,

$$\mathsf{E} \dagger \mathsf{C} = \mathsf{D}.$$

Now put $E = \mathrm{COT}(\mathsf{E})$ and the theorem follows for this case since we are using the identity (on R) as our recursive isomorphism.
Case 2. If $\mathrm{D} \cap \mathrm{A} = \emptyset$, then $B = 0$ or $\mathrm{B} \cap \mathrm{C} \neq \emptyset$. In the former case put $E = 0$ (since A, B, C, D are then all 0) and in the latter case the proof of the existence of an E such that $A + E = B$ and $C = E + D$ is merely a literal variant of that for case 1. This completes the proof.

2.3.3 Corollary. If $E + A_0 + \cdots + A_n + F = B_0 + \cdots + B_n$ then there exist C, D such that for some l, m, $0 \leq l, m \leq n$

$$B_l = C + A_m + D.$$

Proof. The assertion is trivial if $n = 0$. Suppose it holds for all numbers less than n, the hypothesis is true but $(\exists C, D)(B_l = C + A_m + D)$ is false for all $l, m \leq n$. Let $\mathsf{A}_i \in A_i$ and $\mathrm{A}_i = \mathrm{C}‘\mathsf{A}_i$, etc. Then $\mathrm{A}_0 \nsubseteq \mathrm{B}_0$ by assumption so by the proof of the preceding theorem $B_1 + \cdots + B_n = D' + A_1 + \cdots + A_n + F$ for some D'. The result now follows at once by the induction hypothesis.

2.4 *Notation*: $\mathsf{A}^* = \{\langle y, x\rangle : \langle x, y\rangle \in \mathsf{A}\}$, $A^* = \mathrm{COT}(\mathsf{A}^*)$ for any $\mathsf{A} \in A$.

2.4.1 Definition. A^* (A^*) is said to be the *converse* of A (A).

2.4.2 Theorem. (i) $(A + B)^* = B^* + A^*$,
(ii) $A^{**} = A$.

Proof. Left to the reader.

2.4.3 Definition. $A +^* B = (A^* + B^*)^*$.

2.4.4 Theorem. $A +^* B = B + A$.

Proof. Left to the reader.

As a consequence of theorems 2.4.2 and 2.4.4 we have the following *principle of duality*: If P is any property of C.O.T.s involving only $+$, $*$ and $+^*$, variables for C.O.T.s and logical connectives then P is true for C.O.T.s if, and only if, P* is, where the description of P* is obtained from that of P by replacing $+$, $+^*$ by $+^*$, $+$ respectively. We call P* the *dual* of P.

We now introduce our orderings of C.O.T.s.

2.4.5 DEFINITION. A is said to be an *initial (final) segment* of B if

$$x \in C`\mathsf{A} \;\&\; y \in C`\mathsf{B} - C`\mathsf{A} \Rightarrow \langle x, y \rangle \in \mathsf{B} \, (\Rightarrow \langle y, x \rangle \in \mathsf{B}).$$

In such a case we write

$$\mathsf{A} \leq \mathsf{B} \quad (\mathsf{A} \leq^* \mathsf{B}).$$

We also write

$$A \leq B \quad (A \leq^* B),$$

if there exist $\mathsf{A} \in A$ and $\mathsf{B} \in B$ such that $\mathsf{A} \leq \mathsf{B}$ $(\mathsf{A} \leq^* \mathsf{B})$.

2.4.6 LEMMA. $A \leq B$ $(A \leq^* B)$ if, and only if, given $\mathsf{A} \in A$ and $\mathsf{B} \in B$ there is a recursive isomorphism from A onto an initial (final) segment of B.

PROOF. Left to the reader.

2.4.7 THEOREM. (i) $(\exists C)\,(A + C = B) \Rightarrow A \leq B$,
(ii) $(\exists C)\,(C + A = B) \Rightarrow A \leq^* B$.

The converse is false in general and this will be shown to be so later (chapter 14).

The next theorem is very useful indeed. First we need a lemma. Notice that we do not require the A, B, C in the lemma to be separable, merely disjoint.

2.4.8 LEMMA. (M. MORLEY'S LEMMA). If $p: \mathsf{B} \mathbin{\hat{+}} \mathsf{A} \mathbin{\hat{+}} \mathsf{C} \simeq \mathsf{A}$, then

$$\mathsf{B} \mathbin{\hat{+}} \mathsf{A} \simeq \mathsf{A} \simeq \mathsf{A} \mathbin{\hat{+}} \mathsf{C}.$$

PROOF. We assume $\mathsf{B} \neq \emptyset \neq \mathsf{C}$ since otherwise the assertion is trivial. We also assume, without loss of generality, that $\mathsf{A} \subseteq \mathsf{R}$ and $\mathsf{B} \mathbin{\hat{+}} \mathsf{A} \mathbin{\hat{+}} \mathsf{C} \subseteq \mathsf{R}$.

By assumption there exist $b \in \mathrm{B}$ $(= \mathrm{C}`\mathsf{B})$, $c \in \mathrm{C}$. Let $\{x_n\}_{n=0}^{\infty}$ be a recursive enumeration of $\{y : b \preccurlyeq y \preccurlyeq c\}$. We define $q : \mathsf{B} \mp \mathsf{A} \mp \mathsf{C} \simeq \mathsf{B} \mp \mathsf{A}$ as follows:

$$\begin{aligned} q(x, n) &= x \quad && \text{if} \quad x \prec b, \\ &= p(x) && \text{if} \quad c \prec x, \\ &= p(x) && \text{if} \quad x = x_n \;\&\; (\forall m)\,(m \le n \;\&\; x_m \succcurlyeq x_n \Rightarrow x_m \succ p(x_m)), \\ &= x && \text{if} \quad x = x_n \;\&\; (\exists m)\,(m \le n \;\&\; x_m \succcurlyeq x_n \;\&\; p(x) \succcurlyeq x), \end{aligned}$$

is undefined otherwise;

$$q(x) = \lim_{n \to \infty} q(x, n).$$

Clearly q is partial recursive.

If $a \prec x \preccurlyeq c$ for all $a \in \mathrm{A}$ then if $p(x)$ is defined $p(x) \preccurlyeq p(c) \in \mathrm{A}$ and therefore $p(x) \prec x$. So $q(x) = p(x)$ for $x \in \mathrm{C}$. (*)

Hence q maps C onto a final segment of $\mathsf{B} \mp \mathsf{A}$ in a one-one, order preserving fashion. Moreover, if $q(x) = p(x)$ and $x \prec y$, then $q(y) = p(y)$ by clauses 2, 3 of the definition of $q(x, n)$.

Now $p(b) \succ b$ and hence $q(x) = x$ for some x by clause 4. Moreover, if $q(x) = x$ and $y \prec x$, then $q(y) = y$ by clauses 1 and 4.

$$\text{Thus} \quad \mathsf{B} \mp \mathsf{A} = \mathsf{D}_1 \mp \mathsf{D}_2, \qquad (^{**})$$

where $\mathrm{D}_1 = \{x : q^{-1}(x) = x\}$, $\mathrm{D}_2 = \{x : q^{-1}(x) = p^{-1}(x)\}$ and $\mathrm{D}_1 \cap \mathrm{D}_2 = \emptyset$. But then it follows that q maps $\mathsf{B} \mp \mathsf{A} \mp \mathsf{C}$ one-one onto $\mathsf{B} \mp \mathsf{A}$. It remains to prove that q is order preserving. This is clear except when $x \prec y$, $q(x) = x$ and $q(y) = p(y)$. If $x \prec c \preccurlyeq y$ then since q maps C onto a final segment of $\mathsf{B} \mp \mathsf{A}$, $q(x) \prec q(y)$. If $x \prec y \prec c$, then by clause 4, $q(y) = p(y)$ implies there exists $z \preccurlyeq y$ such that $p(z) \succ z$ where by (**) $x \prec z \preccurlyeq y$. Therefore $q(y) = p(y) \succcurlyeq p(z) \succcurlyeq z \succ x = q(x)$. This completes the proof that $q : \mathsf{B} \mp \mathsf{A} \mp \mathsf{C} \simeq \mathsf{B} \mp \mathsf{A}$ and the rest of the lemma follows by symmetry.

The next theorem may be regarded as a recursive analogue of the following theorem attributed to Lindenbaum and Tarski (given in SIERPINSKI, 1958, p. 248).

> "If an order type A is an initial segment of an order type B and the order type B is a final segment of the order type A, then $A = B$."

2.4.9 THEOREM. If $A \le B$ and $B \le^* A$, then $A = B$.

PROOF. By hypothesis, if $\mathsf{B} \in B$ there exist $\mathsf{A} \in A$ and C such that $\mathsf{A} \dotplus \mathsf{C} = \mathsf{B}$ (by the proof of the separation lemma 2.3.1) and similarly there exist $\mathsf{B}' \in B$ and D such that $\mathsf{A} \simeq \mathsf{D} \dotplus \mathsf{B}'$ where $p{:}\,\mathsf{B} \simeq \mathsf{B}'$. Hence $\mathsf{A} \simeq \mathsf{D} \dotplus p(\mathsf{A}) \dotplus p(\mathsf{C})$ and therefore

$$p(\mathsf{A}) \simeq \mathsf{D} \dotplus p(\mathsf{A}) \dotplus p(\mathsf{C}),$$

since $\delta p \supseteq \mathrm{A}$. By Morley's lemma we therefore have

$$p(\mathsf{A}) \simeq p(\mathsf{A}) \dotplus p(\mathsf{C}) = p(\mathsf{B})$$

and the result follows by taking C.O.T.s.

2.4.10 DEFINITION. $A \cdot 0 = 0$, $A \cdot (n+1) = A \cdot n + A$.

2.4.11 THEOREM. (i) If $m > 0$ and $A \cdot m \leq B \cdot m$, then $A \leq B$,
(ii) If $m > 0$ and $A \cdot m \leq^* B \cdot m$, then $A \leq^* B$.

PROOF. By duality it suffices to prove (i). If $m = 1$ there is nothing to prove. Otherwise

$$A \cdot (m+1) = A \cdot m + A \leq B \cdot m + B = B \cdot (m+1),$$

and therefore by the proof of the directed refinement theorem (2.3.2)

$$A \cdot m \leq B \cdot m,$$

or there exist $\mathsf{B} \in B$, $\mathsf{A} \in A$ and a C such that $A \cdot m \geq B \cdot m$ and $\mathsf{A} \dotplus \mathsf{C} \leq^* \mathsf{B}$. In the former case the result follows by the induction hypothesis. In the latter the induction hypothesis yields $B \leq A$. Hence there exist D and $\mathsf{A}(\in A)$ such that $\mathsf{B} \dotplus \mathsf{D} = \mathsf{A}$. By theorem 2.4.9 we obtain $A = B$.

2.4.12 COROLLARY. If $A \cdot m = B \cdot m$, and $m > 0$, then $A = B$.

PROOF. Immediate from theorems 2.4.9 and 2.4.11.

CHAPTER 3

QUORDS

3.1 We now commence our study of proper subcollections of $\mathscr{R}$. Classically, a linear ordering A is said to be a *well-ordering* if either of the equivalent conditions (va) or (vb) below holds.[31]

(va) $(\forall f)(\exists n)(\langle f(n+1), f(n)\rangle \in \mathsf{A} \Rightarrow f(n) = f(n+1))$.

"There are no descending chains."

(vb) $(\forall \mathrm{B})(\emptyset \neq \mathrm{B} \subseteq \mathrm{C}`\mathsf{A} \Rightarrow (\exists b)(b \in \mathrm{B} \,\&\, (\forall a)(a \in \mathrm{B} \Rightarrow \langle b, a\rangle \in \mathsf{A})))$.

"Every non-empty subset has a least element."

In the sequel we shall principally be concerned with such well-orderings but first of all we consider recursive analogues of (va) and (vb).

3.1.1 DEFINITION. A linear ordering A is said to be a *quasi-well-ordering of type* (*a*) (*type* (*b*)) if it satisfies (va) above with the quantifier $(\forall f)$ ranging over *recursive* functions only (if it satisfies (vb) above with the quantifier $(\forall \mathrm{B})$ ranging over r.e. sets only).

3.1.2 LEMMA. (a) If B is a quasi-well-ordering of type (*a*) and $\mathsf{A} \simeq \mathsf{B}$, then A is a quasi-well-ordering of type (*a*).
(b) Similarly for type (*b*).

PROOF. (a) Suppose $p: \mathsf{A} \simeq \mathsf{B}$ and $\{f(n)\}_{n=0}^{\infty}$ is a recursive descending chain in A. Then *pf* is a partial recursive function which is everywhere

defined and hence is recursive. Further p is order preserving, hence $\{pf(n)\}_{n=0}^{\infty}$ is a recursive descending chain in B.
(b) If S is a r.e. set then $p(\mathrm{S})$ is r.e. for partial recursive p. Hence if S is a r.e. subset of C'A without first element then $p(\mathrm{S})$ is a r.e. subset of C'B without first element when $p: \mathsf{A} \simeq \mathsf{B}$.

3.1.3 THEOREM. A linear ordering A is a quasi-well-ordering of type (*a*) if, and only if, it is a quasi-well-ordering of type (*b*).

PROOF. Since we are only dealing with linear orderings which can be embedded in R (or equivalently in some r.e. dense linear ordering) it follows from lemma 3.1.2 and the symmetry of $\simeq$ that we may assume $\mathsf{A} \subseteq \mathsf{R}$. Clearly if A contains no r.e. subset without a first element then A contains no recursive descending chain since $\{f(n)\}$ as a *set* is r.e. for recursive *f*.

Conversely, suppose A contains an infinite r.e. subset S without first element. Let $f(0)$ be an arbitrary element of S and define

$$f(n+1) = \mu_s \{s \in S \;\&\; s \prec f(n) \;\&\; s > f(n)\}\,. \tag{1}$$

Then $f(n)$ is defined for all n since S has no least element under $\preccurlyeq$ and is clearly partial recursive. Hence $\{f(n)\}_{n=0}^{\infty}$ is a recursive descending chain in A. This completes the proof.

This theorem was first proved in a slightly different form by PARIKH (1962, 1966).

3.1.4 DEFINITION. A linear ordering is a quasi-well-ordering if it is a quasi-well-ordering of type (*a*) or type (*b*).

We observe that in definition 3.1.1 for type (*b*) we could have restricted our attention to recursive rather than just r.e. sets, for the subset of S enumerated by *f* in (1) above is a r.e. set of natural numbers enumerated in order of magnitude by a recursive function and hence is recursive. (The other way round is trivial since every recursive set is r.e..)

3.1.5 THEOREM. There exist *recursive* quasi-well-orderings which are not well-orderings.

PROOF. Conditions (i)–(iv) (§ 1.1) and (va) are clearly arithmetic conditions, hence the predicate $\mathfrak{Q}(x)$ (x is (a Gödel number of) a r.e.

quasi-well-ordering) is arithmetic. By SPECTOR (1955, theorem 1), the predicate $\mathfrak{W}(x)$: x is (a Gödel number of) a r.e. well-ordering is a complete predicate of the form

$$(\forall f)(\exists y) R(x, y, f),$$

where $(\forall f)$ ranges over all functions of one variable. Hence by KLEENE (1955) $\mathfrak{W}(x) \Rightarrow \mathfrak{Q}(x)$, but not conversely. Finally, by theorem 1.2.1 and lemma 3.1.2 there is a recursive quasi-well-ordering.

We shall give a direct proof of this result in chapter 14 and it also follows from theorem 7.2.1. That there exist non-recursive quasi-well-orderings is trivial. Let A be an immune set and let A be defined by $\langle x, y\rangle \in \mathsf{A}$ if, and only if $x, y \in \mathrm{A}$ & $y \leq x$; then A is not a well-ordering since A is of type ω^*. But since A contains no infinite r.e. subset, A can contain no recursive descending chain. (Clearly, A is embeddable in R since $\leq$ is recursive.)

3.2.

3.2.1 DEFINITION. A C.O.T. A is said to be a *(recursive) quord* if there is an $\mathsf{A} \in A$ which is a (recursive) quasi-well-ordering. We write $\mathscr{Q}$ for the collection of all quords and $\mathscr{R}\mathscr{Q}$ for the collection of all recursive quords.

3.2.2 THEOREM. A is a (recursive) quord if, and only if, every $\mathsf{A} \in A$ is a (r.e.) quasi-well-ordering.

PROOF. Immediate from lemma 3.1.2, theorem 1.2.1 and the fact that if A is recursive and p is a recursive isomorphism then $p(\mathsf{A})$ is r.e..

Quords, and in particular recursive quords, possess many of the additive and multiplicative properties of C.O.T.s of well-orderings; we shall therefore develop part of this theory before restricting ourselves entirely to C.O.T.s of well-orderings. We note, however, and shall later give the proof that, unlike well-orderings, quasi-well-orderings are not closed under exponentiation (PARIKH, 1962, 1966).

3.2.3 THEOREM. (i) $0 \in \mathscr{Q}$,
(ii) $A, B \in \mathscr{Q} \Rightarrow A + B \in \mathscr{Q}$,
(iii) $A = B + C$ & $A \in \mathscr{Q} \Rightarrow B, C \in \mathscr{Q}$.

PROOF. Left to the reader.

It is often useful to consider descending chains of the form $\{f^i(a)\}$ so we give the following definition.[32]

3.2.4 Definition. A recursive infinite descending chain $\{g(n)\}_{n=0}^{\infty}$ is said to be a *splinter* if there is a one-one partial recursive function, f, such that $g(0)=c$, $g(n+1)=f(g(n))$ where c is some fixed number. We write $g(n)=f^n(c)$.

3.2.5 Theorem. A linear ordering is a quasi-well-ordering if, and only if, it contains no splinter.

Proof. Let A be a linear ordering and suppose that $\{g(n)\}_{n=0}^{\infty}$ is a recursive descending chain in A. Define f, c as follows:

$$c = g(0), \quad f(n+1) = g(1 + \mu_y\{g(y) = n+1\})$$

(cf. Ullian, 1960, p. 33). Then $\{g(n)\}$ is a splinter in A.

Conversely, suppose that $\{f^n(c)\}$ is a splinter in the linear ordering A. Let g be the function defined by

$$g(0) = c, \quad g(n+1) = f(g(n)).$$

Then g is totally defined and computable, hence recursive; i.e. $\{g(n)\}$ is a recursive descending chain in A. This completes the proof.

The next lemma is essentially proved (in a different context) by Sierpinski (1958, p. 264).

3.2.6 Lemma. If A is a quasi-well-ordering and f is a one-one, partial recursive, order preserving map of $C`\mathsf{A}$ into $C`\mathsf{A}$ then

$$\langle x, f(x)\rangle \in \mathsf{A}$$

for all $x \in C`\mathsf{A}$.

Proof. Since $x \in C`\mathsf{A} \Rightarrow f(x) \in C`\mathsf{A}$ and A is linear either $\langle x, f(x)\rangle \in \mathsf{A}$ or $\langle f(x), x\rangle \in \mathsf{A}$ (or both). Suppose $\langle f(x), x\rangle \in \mathsf{A}$ and $\langle x, f(x)\rangle \notin \mathsf{A}$ then, since f is order preserving we easily establish by induction that for all n $\langle f^{n+1}(x), f^n(x)\rangle \in \mathsf{A}$, i.e. $\{f^n(x)\}_{n=0}^{\infty}$ is a splinter in A. The conclusion now follows by theorem 3.2.5.

3.2.7 Theorem. If A is a quord and $A=B+A+C$ then $C=0$.

Proof. Assume the hypothesis and $C \neq 0$. Then there exist quasi-well-orderings $\mathsf{A}\in A$, $\mathsf{B}\in B$ and $\mathsf{C}\in C$ and a recursive isomorphism f such that

$$f\colon \mathsf{B} \mathbin{\hat{+}} \mathsf{A} \mathbin{\hat{+}} \mathsf{C} \simeq \mathsf{A}.$$

By lemma 3.2.6 if $x\in C'\mathsf{A}$, $\langle x, f(x)\rangle\in\mathsf{A}$. Since $\mathsf{C}\neq\emptyset$, $\exists x\in C'\mathsf{A}$ with $f^{-1}(x)\in C'\mathsf{C}$ hence $\langle x, f^{-1}(x)\rangle\in\mathsf{B}\mathbin{\hat{+}}\mathsf{A}\mathbin{\hat{+}}\mathsf{C}$ and $\langle f(x), x\rangle\in\mathsf{A}$, $x\neq f(x)$. But this implies $\langle x, f(x)\rangle\notin\mathsf{A}$ which is a contradiction so we conclude $\mathsf{C}=\emptyset$ whence $C=0$.

3.2.8 Corollary. If A is a quord, then $A+B=A$ implies $B=0$.

3.2.9 Corollary. If A is a quord, then

$$A+B=A+C \Leftrightarrow B=C.$$

Proof. Suppose $A+B=A+C$, then by the directed refinement theorem (2.3.2) there is an E such that either $A=A+E$ and $E+B=C$ or $A+E=A$ and $B=E+C$. In either case, by corollary 3.2.8, $E=0$ and $B=C$. The converse is trivial.

3.3 We know very little about the order structure of recursive quasi-well-orderings and have only the following result. But we shall show in chapter 15 that for each countably infinite order type there exist $2^{\aleph_0}$ non-recursive quasi-well-orderings of that type which are not recursively isomorphic.

3.3.1 Theorem. If A is a r.e. quasi-well-ordering then the order type of A is either finite or of the form

$$\omega + \tau,$$

where τ is some (countable) order type.

Proof. Since A is r.e.

$$\mathsf{A} \restriction a = \{\langle x, y\rangle : \langle a, x\rangle\in\mathsf{A} \;\&\; \langle x, y\rangle\in\mathsf{A} \;\&\; a \neq x\}$$

is also r.e. for every $a\in C'\mathsf{A}$. Hence by lemma 2.1.2 each $\mathsf{A}\restriction a$ has a least element and hence every element in $C'\mathsf{A}$ has an immediate successor if it has any element succeeding it. Similarly, A has a first element since $C'\mathsf{A}$ is r.e. The theorem now follows at once.

3.3.2 Corollary. If A is a recursive quord, then every $\mathsf{A}\in A$ has a least element.

CHAPTER 4

THE ORDERINGS

4.1 We denote the classical denumerable ordinals (and variables ranging over the denumerable classical ordinals) by lower case Greek letters $\alpha, \beta, \gamma, \ldots$. We regard classical ordinals as order types which contain a sub-ordering of R which is a well-ordering. Clearly any well-ordering whose type is a denumerable ordinal can be embedded by a (possibly non-recursive) isomorphism in R; but we shall exhibit a well-ordering (of type $\omega \cdot 2$) which cannot be embedded in R by a recursive isomorphism.

4.1.1 DEFINITION. If A is the C.O.T. of a (recursive) well-ordering, then A is said to be a *(recursive) co-ordinal*. We let $\mathscr{C}$ ($\mathscr{RC}$) denote the collection of all co-ordinals (of all recursive co-ordinals). If α is the classical ordinal such that $A \subseteq \alpha$ then α is said to be the (classical) ordinal of A and of any $\mathsf{A} \in A$ and we write

$$|A| = \alpha \quad \text{and} \quad |\mathsf{A}| = \alpha\,.$$

4.1.2 COROLLARY. $A \subseteq |A|$.

4.1.3 THEOREM. A is a (recursive) co-ordinal if, and only if, every $\mathsf{A} \in A$ is a (r.e.) well-ordering.

PROOF. Left to the reader. (Use theorem 1.2.1.)

4.1.4 THEOREM. Any two linear orderings with fields of the same finite cardinal are recursive well-orderings which are recursively isomorphic by a recursive permutation.

PROOF. That finite linear orderings are well-orderings is well-known. Any two finite well-ordered sets of the same cardinal can be mapped onto each other in a one-one order preserving manner by a recursive permutation since they can be so mapped by some permutation and every permutation of the integers which interchanges only finitely many numbers is necessarily recursive. It follows that any two finite linear orderings with fields of the same cardinal are recursively isomorphic by a recursive permutation.

4.1.5 COROLLARY. If A is finite (i.e. if $\mathsf{A} \in A \Rightarrow \mathsf{C}`\mathsf{A}$ is finite), then $|A| = A$.

We denote finite co-ordinals by 0, 1, 2, ... in the obvious way since it will be clear from the context whether n denotes a co-ordinal or a natural number. We also use $\mathscr{N}$ to denote the collection of all finite co-ordinals. We observe that our notation hitherto has been consistent with this usage.

4.1.6 THEOREM. (i) $0 \in \mathscr{C}$,
(ii) $A, B \in \mathscr{C} \Rightarrow A + B \in \mathscr{C}$,
(iii) $A = B + C \;\&\; A \in \mathscr{C} \Rightarrow B, C \in \mathscr{C}$.

PROOF. Left to the reader.

4.1.7 LEMMA. $|A+B| = |A| + |B|$.

4.1.8 THEOREM. (i)

$$\begin{array}{ccccccc} \mathscr{N} & \subset & \mathscr{R}\mathscr{C} & \subset & \mathscr{R}\mathscr{Q} & \subset & \mathscr{R} \\ & & \cap & & \cap & & \\ & & \mathscr{C} & \subset & \mathscr{Q} & & \end{array}$$

(ii) The cardinalities of $\mathscr{N}$, $\mathscr{R}\mathscr{C}$, $\mathscr{R}\mathscr{Q}$ are all $\aleph_0$ and of $\mathscr{C}$, $\mathscr{Q}$, $\mathscr{R}$ are c (the cardinal of the continuum).

PROOF. (i) Every finite sub-ordering of R is a recursive well-ordering hence $\mathscr{N} \subseteq \mathscr{R}\mathscr{C} \subseteq \mathscr{C}$. $\mathscr{N} \neq \mathscr{R}\mathscr{C}$ since there exist infinite recursive well-orderings and $\mathscr{R}\mathscr{C} \neq \mathscr{C}$ by part (ii) below. Since every co-ordinal is a quord, by theorem 3.1.5 we have $\mathscr{R}\mathscr{C} \subset \mathscr{R}\mathscr{Q}$, and by theorem 1.2.3, $\mathscr{R}\mathscr{Q} \subseteq \mathscr{R}$. $\mathscr{R}\mathscr{Q} \neq \mathscr{R}$ and $\mathscr{R}\mathscr{Q} \neq \mathscr{Q}$ by part (ii). Finally $\mathscr{C} \subseteq \mathscr{Q}$, clearly, and $\mathscr{C} \neq \mathscr{Q}$ by the example following theorem 3.1.5.

(ii) Since $\mathscr{N}$, $\mathscr{R}\mathscr{C}$, $\mathscr{R}\mathscr{Q}$ contain only equivalence classes of r.e. orderings they are at most countable. Since the cardinal of $\mathscr{N}$ is $\aleph_0$, they all have cardinal $\aleph_0$.

There exist c distinct sub-orderings of Seq. Since there are only $\aleph_0$

recursive isomorphisms these sub-orderings lie in x classes where $\aleph_0 \cdot x = c$. Using the axiom of choice we conclude $x = c$.

4.2.

4.2.1 THEOREM. If $B \in \mathscr{C}$ then

$$A \le B \Leftrightarrow (\exists C)(A + C = B),$$

and

$$A \le^* B \Leftrightarrow (\exists C)(C + A = B).$$

PROOF. Immediate from theorem 2.1.2.

4.2.2 THEOREM. If B is a co-ordinal then the following statements are equivalent: (i) $A \le B \;\&\; A \ne B$,

(ii) there is a (unique) co-ordinal $C \ne 0$ such that $A + C = B$.

PROOF. Suppose $A + C_1 = A + C_2 = B$, then by corollary 3.2.9, $C_1 = C_2$. By theorem 4.2.1 there is a C such that $A + C = B$ and since $A \ne B$ we must have $C \ne 0$. The converse is trivial.

4.2.3 THEOREM.
(i) $A \le A$, (i)* $A \le^* A$,
(ii) $0 \le A$, (ii)* $0 \le^* A$,
(iii) $A \le 0 \Leftrightarrow A = 0$, (iii)* $A \le^* 0 \Leftrightarrow A = 0$,
(iv) $A \le B \,\&\, B \le C \Rightarrow A \le C$, (iv)* $A \le^* B \,\&\, B \le^* C \Rightarrow A \le^* C$,
(v) $B \le C \Rightarrow A + B \le A + C$, (v)* $A \le^* B \Rightarrow A + C \le^* B + C$.

PROOF. Left to the reader.

4.2.4 THEOREM. If A is a quord, then
(i) $A \not< A$ (i)* $A \not<^* A$,
(ii) $A \le B \,\&\, B \le A \Rightarrow A = B$, (ii)* $A \le^* B \,\&\, B \le^* A \Rightarrow A = B$.

PROOF. Immediate from lemma 2.4.6 and lemma 3.2.6.

We shall now write $A < B$ for $A \le B \,\&\, A \ne B$ when B is a quord and $\mathsf{A} < \mathsf{B}$ for $\mathsf{A} \le \mathsf{B}$ and $\mathsf{A} \ne \mathsf{B}$.

4.2.5 LEMMA. If B is a quord, $\mathsf{A} \le \mathsf{B} \in B$ and $A = \mathrm{COT}(\mathsf{A})$, then $A < B$.
PROOF. Left to the reader.

We now give an easy proof of 2.4.9 for quords.

4.2.6 THEOREM. If A or B is a quord, then

$$A \leq B \ \& \ B \leq^* A \Rightarrow A = B .$$

PROOF. Assume the hypotheses and that $A \neq B$, then there exist $\mathsf{A}, \mathsf{A}' \in A$, $\mathsf{B} \in B$ and recursive isomorphisms p, q, r such that

$$p : \mathsf{A} \simeq \mathsf{B}' \leq \mathsf{B}, \tag{1}$$

$$q : \mathsf{B} \simeq \mathsf{A}'' \leq^* \mathsf{A}' \tag{2}$$

and

$$r : \mathsf{A} \simeq \mathsf{A}' ,$$

where both of the inequalities (1), (2) are strict since $A \neq B$. Hence there exist $\mathsf{C}, \mathsf{D} \neq \emptyset$ such that

$$\mathsf{A} = \mathsf{C} \dotplus r^{-1} qp(\mathsf{A}) \dotplus \mathsf{D} .$$

But this is impossible by the proof of theorem 3.2.7 and we conclude $A = B$.

4.2.7 DEFINITION. A partial ordering $\leq$ is a *tree ordering* if

$$A, B \leq C \Rightarrow A \leq B \quad \text{or} \quad B \leq A .$$

It is a *rooted* tree ordering if it is a tree ordering and there is an element 0 such that $0 \leq A$ for every A. A *partial well-ordering* is a relation $\leq$ satisfying (i), (ii), (iii) (§ 1.1) and (va) (§ 3.1).

The following theorem sums up all the important properties of the orderings that we have so far obtained and adds some more.

4.2.8 THEOREM. (i) $\leq$ is a rooted tree partial well-ordering of $\mathscr{C}$,
(ii) $\leq^*$ is a tree ordering of $\mathscr{Q}$ (and hence of $\mathscr{C}$),
(iii) $\leq$ is a rooted tree ordering but not a partial well-ordering of $\mathscr{Q}$.

PROOF. (i) Since the (classical) ordinals are well-ordered it follows from lemma 4.1.7 and theorem 4.2.3 that there are no descending chains under $\leq$. The other required properties follow immediately from theorems 4.2.3 (i), 4.2.4 (ii), 4.2.3 (iv). By theorem 4.2.1 and theorem 2.3.2 (directed refinement theorem) it follows at once that $\leq$ is a tree ordering and from theorem 4.2.3 (ii) that $\leq$ is a rooted tree ordering.

We leave the details of the proofs of (ii) and (iii) to the reader except for the proof that $\leq$ is not a partial well-ordering of $\mathscr{Q}$. Let A be a recursive quord which is not a C.O.T. of a well-ordering[41] and let $\{a(n)\}_{n=0}^{\infty}$ be an infinite (necessarily non-recursive) descending chain in an $\mathsf{A} \in A$ such that $\mathsf{A} \subseteq \mathsf{R}$. Let

$$\mathsf{A}_n = \mathsf{A}\left[\{x : x \preccurlyeq a(n)\}\right],$$

then, clearly, each A_n determines a quord A_n and by theorem 2.1.2 we easily obtain

$$m < n \Rightarrow A_m > A_n .$$

4.2.9 THEOREM. If $C \in \mathscr{R}\mathscr{Q}$ then

$$C \leq^* A \Leftrightarrow (\exists B)(B + C = A).$$

PROOF. Since $A \subseteq R$ there is an $\mathsf{A} \in A$ such that $\mathsf{A} \subseteq \mathsf{R}$. Further since $C \leq^* A$ there is a $\mathsf{C} \in C$ such that

$$\mathsf{C} \leq^* \mathsf{A} \subseteq \mathsf{R}.$$

By theorem 3.3.1, C has a least element c_0 or is empty. In the latter case take $B = A$; in the former case let

$$\mathsf{B} = \mathsf{A}\left[\{x : x \preccurlyeq c_0\}\right];$$

then by theorem 2.1.2,

$$\mathsf{B})(\mathsf{C} \quad \text{and clearly} \quad \mathsf{B} \mathbin{\dot{+}} \mathsf{C} = \mathsf{A}.$$

Taking C.O.T.s yields the required result. The converse is trivial.

4.3 We know that (for quords) $A = B$ is equivalent both to

$$A \leq B \;\&\; B \leq A$$

and to

$$A \leq B \;\&\; B \leq^* A,$$

but we shall later show, using the fact that there exist incomparable co-ordinals, that $\leq^*$ is not anti-symmetric even for co-ordinals. One might wonder whether the analogue of the classical theorem

> "If a well-ordered set A is similar to a subset of a well-ordered set B, then A is similar to an initial segment of B"

is true for quords or co-ordinals. However, we showed in CROSSLEY (1965, Appendix) that the analogue is not true for co-ordinals.

Finally, we give an example of a well-ordering of type $\omega \cdot 2$ which cannot be embedded in R by a recursive isomorphism.

4.3.1 EXAMPLE. Let S be a r.e. non-recursive set; then $\bar{S}$ is infinite. Let U be the well-ordering of type $\omega \cdot 2$ defined by

$$\begin{aligned} \langle x, y\rangle \in \mathsf{U} \Leftrightarrow & \; x \in \mathrm{S} \;\&\; y \in \bar{\mathrm{S}} \\ & \vee \; x, y \in \mathrm{S} \;\&\; x \leq y \\ & \vee \; x, y \in \bar{\mathrm{S}} \;\&\; x \leq y . \end{aligned}$$

Further let r be the least element of $\bar{S}$. Suppose

$$p : \mathsf{U} \simeq \mathsf{U}' \subseteq \mathsf{R},$$

then the initial ω segment of $\mathsf{U}' \subseteq \{x : x \prec p(r)\} = \mathrm{A}$, say, and the final segment of type ω is contained in $\{x : x \succcurlyeq p(r)\} = \mathrm{B}$, say. Hence

$$\mathrm{S} \subseteq p^{-1}(\mathrm{A}) \quad \text{and} \quad \bar{\mathrm{S}} \subseteq p^{-1}(\mathrm{B}),$$

i.e. S and $\bar{S}$ are contained in disjoint r.e. sets. This contradicts the fact that $\bar{S}$ is not recursive and we conclude that there can be no such recursive isomorphism p.

CHAPTER 5

CO-ORDINALS

5.1 We have already introduced co-ordinals and in this chapter we investigate their basic properties.

5.1.1 DEFINITION. A co-ordinal A is said to be a *successor number* if

$$(\exists B)\,(A = B + 1)$$

and a *limit number* if it is neither a successor number nor zero.

5.1.2 THEOREM. If A is a successor number (limit number), then $|A|$ is a successor number (limit number).

PROOF. Obvious.

If $A = B + C$, then we put $C = A - B$ (when A, B, C are quords). By corollary 3.2.9, if $A - B$ exists it is unique.

5.1.3 THEOREM. (i) $A - A = 0$,

(ii) $(A + B) - A = B$,

(iii) if A is a co-ordinal (or a quord), then $B + (A - B) = A$,

(iv) if A is a co-ordinal (or a quord) and $B + C \leq A$, then $A - (B + C) = (A - B) - C$.

Proof of (iv).

$$B + C \leq A \Rightarrow (E!\,D)\,(A = B + C + D),$$

hence

$$A - (B + C) = D.$$

Also,

$$A - B = C + D \quad \text{and therefore}$$
$$(A - B) - C = D.$$

5.1.4 THEOREM. $C < A + 1 \Leftrightarrow C \leq A$.

PROOF. Take $\mathsf{A} \in A$ and suppose $C < A + 1$; then there is a $\mathsf{C} \in C$ such that

$$\mathsf{C} < \mathsf{A} + \mathsf{1} \quad \text{where} \quad \mathsf{1} = \{\langle 0, 0 \rangle\} \in 1.$$

Clearly, $\mathsf{C} \leq \hat{0}\mathsf{A}$ and hence $C \leq A$. The converse is trivial.

5.1.5 COROLLARY. If A is a quord, then $A + 1$ is the unique quord B such that

$$A < B \Rightarrow (\forall C)(C < B \Rightarrow C \leq A).$$

By virtue of this corollary we call $A + 1$ *the successor* of A.

5.1.6 THEOREM. If A is a successor number where $|A| = \lambda + n$ and λ is a limit ordinal then the sequence

$$A - n, \ldots, A - 2, A - 1, A, A + 1, A + 2, \ldots$$

of length ω, is uniquely defined.

PROOF. Since A is a successor number $\mathsf{A} \in A$ implies A has a finite final segment. Now every finite set is recursive and hence separable from its complement in C'A, hence $A - 1, A - 2, \ldots, A - m$ are well-defined provided $A - (m - 1)$ is a successor number. The rest is clear.

5.2 By the same argument that finite sets are recursive and hence separable from their complements in an ordering we leave the reader to prove

5.2.1 LEMMA. A co-ordinal A is infinite if, and only if, $n < A$ for all n.

5.2.2 THEOREM. A co-ordinal A is a limit number if, and only if,

$$B < A \Rightarrow B + 1 < A.$$

PROOF. Immediate from theorem 5.1.4 and the classical theorem that if β is not a successor number then

$$\alpha < \beta \Rightarrow \alpha + 1 < \beta .$$

5.2.3 COROLLARY. If A is a co-ordinal, then

$$A < B \Leftrightarrow A + 1 \leq B .$$

PROOF. Left to the reader.

We know from theorem 1.1.7 that constructive order types give a finer classification of well-orderings than do (classical) order types and the next theorem gives a measure of this fact. Most of the effort in the latter half of part one of this monograph will be devoted to trying to single out one co-ordinal for each denumerable ordinal, a problem that is not yet solved except for relatively small ordinals.

5.2.4 THEOREM. There are c distinct co-ordinals of ordinal ω.

PROOF. There are c distinct infinite subsets of $\mathscr{N} - \{0\}$. Consider such a subset S and assign to it the sub-ordering S of R given by

$$\langle x, y\rangle \in \mathsf{S} \Leftrightarrow x, y \in \mathrm{S} \ \& \ \langle x\rangle \preccurlyeq \langle y\rangle .$$

Then these c sets are spread among, say, x equivalence classes. Each equivalence class contains at most $\aleph_0$ members since there are only $\aleph_0$ recursive isomorphisms. Hence $\aleph_0 \cdot x = c$ and therefore (using the axiom of choice) $x = c$.

Notation.[51] $\mathsf{W}' = \{\langle\langle i\rangle, \langle j\rangle\rangle : \langle i\rangle \preccurlyeq \langle j\rangle \ \& \, i, j > 0\}$, $W = \mathrm{COT}(\mathsf{W}')$. Let $\mathrm{V} \subseteq \mathscr{N} - \{0\}$ be a fixed but arbitrary r.e. non-recursive set, then we put

$$\mathsf{V} = \{\langle\langle i\rangle, \langle j\rangle\rangle : i, j \in \mathrm{V} \ \& \ i \leq j\}$$

and $V = \mathrm{COT}(\mathsf{V})$.

Clearly,

$$\mathsf{W}' = \{\langle\langle i\rangle, \langle j\rangle\rangle : i \leq j\} \simeq \{\langle i, j\rangle : i \leq j\} = \mathsf{W} .$$

5.2.5 DEFINITION. If $\mathsf{W}'' \in W$ then W'' is said to be a *natural* well-ordering of type ω and W is said to be the *natural* co-ordinal of type ω. We call V, V the *basic counterexamples* of well-orderings, co-ordinals, respectively.[52]

5.2.6 THEOREM. (i) $1+W=W$, (ii) $1+V\neq V$.

PROOF. (i) Take $\mathbf{1}\in 1$ as in the proof of theorem 5.1.4 and let p be the map defined by

$$\begin{aligned} p(x) &= 0 \quad \text{if} \quad x=\hat{0}(0), \\ &= i+1 \quad \text{if} \quad x=\hat{1}(\langle i\rangle), \\ &\text{is undefined otherwise.} \end{aligned}$$

Then it is easy to verify that

$$p:\mathbf{1}+\mathsf{W}'\simeq\mathsf{W}'.$$

(ii) Suppose $1+V=V$. Since V is r.e. non-recursive, $\bar{\mathsf{V}}$ is non-empty, say $a_0\in\bar{\mathsf{V}}$. Therefore there is a recursive isomorphism p such that

$$p:\{\langle a_0, a_0\rangle\}\hat{+}\mathsf{V}\simeq\mathsf{V}.$$

Hence

$$\mathsf{V}=\{\langle p^m(a_0), p^n(a_0)\rangle : m\leq n\}$$

and if g is the recursive isomorphism defined by

$$\begin{aligned} g(x) &= p^n(a_0) \quad \text{if} \quad x=n, \\ &\text{is undefined otherwise,} \end{aligned}$$

then

$$g:\mathsf{W}\simeq\mathsf{V}.$$

But then g enumerates V in order of magnitude and it follows that V is recursive, contradicting our choice of V.

Co-ordinals A, B are said to be *incomparable* if $A\nleq B$ and $B\nleq A$.

5.2.7 COROLLARY. V and W are incomparable.

PROOF. By the theorem $V\neq W$. But $V<W$ or $W<V$ implies V or W is finite, which is impossible.

5.3 If A is a linear ordering then we write

$$x\leq_{\mathsf{A}} y \quad \text{for} \quad \langle x, y\rangle\in\mathsf{A}$$

and

$$x<_{\mathsf{A}} y \quad \text{for} \quad x\leq_{\mathsf{A}} y \;\&\; x\neq y.$$

If $a \in C`\mathsf{A}$, then a determines $\left\{\begin{matrix}\text{an open}\\ \text{a closed}\end{matrix}\right\}$ initial segment

$$\left\{\begin{array}{l} a)\,\mathsf{A} = \{\langle x, y\rangle : x \leq_{\mathsf{A}} y <_{\mathsf{A}} a\} \\ a]\,\mathsf{A} = \{\langle x, y\rangle : x \leq_{\mathsf{A}} y \leq_{\mathsf{A}} a\} \end{array}\right.$$

and $\left\{\begin{matrix}\text{an open}\\ \text{a closed}\end{matrix}\right\}$ final segment

$$\left\{\begin{array}{l} \mathsf{A}\,(a = \{\langle x, y\rangle : a <_{\mathsf{A}} x \leq_{\mathsf{A}} y\} \\ \mathsf{A}\,[a = \{\langle x, y\rangle : a \leq_{\mathsf{A}} x \leq_{\mathsf{A}} y\}\,. \end{array}\right.$$

Clearly, a is the first element of $\mathsf{A}\,[a$.

If A is a well-ordering of ordinal α, then $a)\,\mathsf{A}$ is a well-ordering of ordinal β, and we write

$$\alpha = |\mathsf{A}|\,, \quad \beta = |a)\,\mathsf{A}|\,. \tag{1}$$

Conversely, if $\beta < \alpha$, then there exists $a \in C`\mathsf{A}$ such that (1) holds. We shall also write

$$|a|_{\mathsf{A}} \quad \text{or} \quad |a| \quad \text{for} \quad |a)\,\mathsf{A}|\,.$$

5.3.1 LEMMA. If $|A| = \alpha$ and $\beta < \alpha$ then

$$(\exists B)\,(B < A\ \&\ |B| = \beta)\,.$$

PROOF. Take $\mathsf{A} \in A$, then there exists $a \in C`\mathsf{A}$ such that (1) holds. Let

$$B = \mathrm{COT}\,(a)\mathsf{A})$$

then clearly, $B < A$.

We write $\mathscr{P}(A) = \{B : B < A\}$
and $\mathscr{P}^+(A) = \{B : B \leq A\}$
and we call $\mathscr{P}(A)$ and $\mathscr{P}^+(A)$ *paths*. We let $\mathscr{P}$ denote the collection of all paths of co-ordinals. We shall later show that there exists a path right through the countable ordinals but here we only show that paths "look like" initial segments of the classical ordinals.

5.3.2 THEOREM. If $|A| = \alpha$, then $\mathscr{P}(A)$ is isomorphic to α under the map $|\ |$ and $\mathscr{P}^+(A)$ is isomorphic to $\alpha + 1$ under the same map.

PROOF. Immediate from lemma 5.3.1.

Later we shall extend this theorem to certain paths which are closed under given co-ordinal operations and shall show that the same mapping is an isomorphism with respect to the corresponding classical operations.

The next theorem is useful when we wish to employ classical results to get results about co-ordinals. In fact, in the next section we shall so employ it.

5.3.4 THEOREM. If A, $B \leq C$ and C is a co-ordinal then

$$|A| \leq |B| \Leftrightarrow A \leq B$$

and

$$|A| = |B| \Leftrightarrow A = B.$$

PROOF. By theorem 4.2.8, A and B are comparable. By theorem 5.3.2, since $|\ |$ is an isomorphism on $\mathscr{P}(C)$ to $|C|$, $A \leq B$.

The second assertion follows from the anti-symmetry of $\leq$.

5.3.5 COROLLARY. If A, B are comparable, then

$$|A| < |B| \Leftrightarrow A < B$$

and

$$|A| = |B| \Leftrightarrow A = B.$$

5.3.6 COROLLARY. If $|A| = \alpha$ and $\beta < \alpha$, then

$$(\mathrm{E}!B)(\mathrm{E}!C)(B + C = A \;\&\; |B| = \beta).$$

PROOF. Immediate from lemma 5.3.1, theorem 5.3.4 and theorem 4.2.2.

5.4 This section is devoted to a characterization of commuting co-ordinals. We have no corresponding result for the multiplicative case.

5.4.1 THEOREM. If A, B are co-ordinals then $A + B = B + A$ if, and only if, there exists a co-ordinal C and finite co-ordinals m, n such that

$$A = C \cdot m \quad \text{and} \quad B = C \cdot n.$$

PROOF. Classically we know (*see*, e.g. SIERPINSKI, 1958) that $\alpha + \beta = \beta + \alpha$ if, and only if, there exists γ and finite ordinals m, n such that

$$\alpha = \gamma \cdot m \quad \text{and} \quad \beta = \gamma \cdot n.$$

Clearly we may assume $(m, n) = 1$, i.e. m, n are co-prime.

Now suppose $|A|=\alpha$, $|B|=\beta$ and $A+B=B+A$. If α or β is 0 there is nothing to prove, so we assume $\alpha \geq \beta > 0$ from which it follows that $m \geq n$. By corollary 5.3.6 there exist $C'_1, \ldots, C'_{m-1}, D'_1, \ldots, D'_{n-1}$ such that

$$|C'_i| = \gamma \cdot i, \quad |D'_j| = \gamma \cdot j \quad (i = 1, \ldots, m-1; j = 1, \ldots, n-1),$$

where

$$C'_i < A, \; D'_j < B.$$

Since the C'_i are bounded by A they are all comparable and so by applying corollary 5.3.6 $(m-1)$ times we obtain $C_1, \ldots, C_m$ where

$$C_1 = C'_1, C_2 = C'_2 - C'_1, \ldots, C_m = A - C'_{m-1}.$$

Similarly we obtain $D_1, \ldots, D_n$ since the D'_j are less than B. Set $C_{m+j} = D_j$ for $j=1, \ldots, n$ and $m+n=r$. Then

$$|C_i| = \gamma \quad \text{for} \quad i = 1, \ldots, r,$$

and

$$A + B = C_1 + \cdots + C_m + C_{m+1} + \cdots + C_r = B + A = C_{m+1} + \cdots + C_r + C_1 + \cdots + C_m.$$

Again by corollary 5.3.6 we obtain

$$C_1 = C_{m+1}, C_1 + C_2 = C_{m+1} + C_{m+2}, \ldots \quad C_1 + \cdots + C_{r-1} = C_{m+1} + \cdots + C_{m-1}$$

and therefore by corollary 3.2.9,

$$C_i = C_{\sigma(i)} \quad (i = 1, \ldots, r), \tag{2}$$

where σ is the permutation[53]

$$\begin{pmatrix} 1 & \ldots n & n+1 \ldots r \\ m+1 \ldots r & 1 & \ldots m \end{pmatrix}$$

on r letters.

Let τ be the cycle $(12 \ldots r)$ so that

$$\begin{aligned} \tau(i) &= i+1 \quad \text{if} \quad i < r, \\ &= 1 \qquad\;\; \text{if} \quad i = r, \end{aligned}$$

then $\sigma = \tau^m$ and τ is of order r. Now $(m, r)=1$ so there exist integers u, v such that

$$mu + rv = 1.$$

Hence

$$\sigma^u = \tau^{mu} = \tau^{1-rv} = \tau .$$

It now follows from (2) that

$$\begin{aligned} C_i = C_{\sigma^u(i)} &= C_{i+1} \quad \text{if} \quad i < r, \\ &= C_1 \quad \text{if} \quad i = r, \end{aligned}$$

so $C_i = C_j = C$, say, for $i, j = 1, \ldots, r$ and we conclude

$$A = C \cdot m , \quad B = C \cdot n .$$

The converse is trivial.

CHAPTER 6

MULTIPLICATION

6.1 We define multiplication of orderings in the usual way except that we code the ordered pairs we obtain by means of the function $j(x, y)$ defined by

$$j(x, y) = \tfrac{1}{2}(x + y)(x + y + 1) + x.$$

We recall that there exist k, l such that

$$j(k(x), l(x)) = x$$

and that j, k, l are all primitive recursive.

6.1.1 DEFINITION. $\mathsf{A} \cdot \mathsf{B} = \{\langle j(a, b), j(a', b')\rangle : a, a' \in \mathrm{C}`\mathsf{A}\ \&$
$(b <_{\mathsf{B}} b' \ \vee .\ b = b' \in \mathrm{C}`\mathsf{B}\ \&\ a \leq_{\mathsf{A}} a')\}$.

6.1.2 THEOREM. If A, B are linear orderings (quasi-well-orderings, well-orderings) then $\mathsf{A} \cdot \mathsf{B}$ is a linear ordering (quasi-well-ordering, well-ordering). If A, B are r.e. (recursive), then so too is $\mathsf{A} \cdot \mathsf{B}$.

PROOF. All except the case of quasi-well-orderings follow at once from the classical definition of multiplication of orderings (cf. SIERPINSKI, 1958, p. 229).

Suppose A, B are quasi-well-orderings and that $\mathsf{A} \cdot \mathsf{B}$ is not. Then there is a recursive descending chain in $\mathsf{A} \cdot \mathsf{B}$. Since every element of the field of $\mathsf{A} \cdot \mathsf{B}$ is of the form $j(a, b)$, this chain has the form

$$\{j(a_n, b_n)\}_{n=0}^{\infty},$$

where $a_n \in C\text{‘}\mathsf{A}$ and $b_n \in C\text{‘}\mathsf{B}$. Let $A' = \{a_n : n \geq 0\}$ and $B' = \{b_n : n \geq 0\}$, then there are four cases to consider:

(i) A′ and B′ are both finite,
(ii) A′ is infinite and B′ is finite,
(iii) A′ is finite and B′ is infinite,
(iv) A′ and B′ are both infinite.

(i) is impossible since then $\{j(a_n, b_n)\}$ would contain only finitely many elements.

(ii) Since B′ is finite, there is at least one number $b \in B'$ for which

$$\{j(a_n, b) : a_n \in A'\}$$

is infinite. Let the distinct a_n in this set be $a(n_i)$ $i = 0, 1, \ldots$ where $i < j \Rightarrow n_i < n_j$, then

$$\{j(a(n_i), b)\}_{i=0}^{\infty}$$

is a recursive descending chain in $\mathsf{A} \cdot \mathsf{B}$ since

$$a(n_0) = a_0,$$
$$a(n_{i+1}) = a_r,$$

where

$$r = \mu_s \{s > t .\&. u < s \Rightarrow a_u \neq a_s\} \text{ and } a_t = a(n_i).$$

It follows at once that

$$\{a(n_i)\}_{i=0}^{\infty}$$

is a recursive descending chain in A which is a contradiction.

(iii) This case is dealt with in a manner very similar to (ii). We omit the details.

(iv) Let

$$\{b(n_i) : i \geq 0\}$$

be the set of distinct b_i where $i < j \Rightarrow n_i < n_j$. Then every $b(n_i)$ occurs in $\{b_n\}_{n=0}^{\infty}$ at most finitely many times for the following two reasons:

1. If $b_j = b(n_i)$ for some fixed i and all j greater than some j_0, then there are only finitely many distinct b_n, namely those occurring in

$$b_0, \ldots, b_{j_0}.$$

2. If $i < j$ and

$$\langle b(n_i), b(n_j) \rangle \in \mathsf{B}$$

then, since

$$\langle b(n_j), b(n_i) \rangle \in \mathsf{B}$$

by the definition of $\mathsf{A}\cdot\mathsf{B}$ and since B is anti-symmetric,

$$b(n_i) = b(n_j),$$

which contradicts our assumption that $i<j$. Hence $\{b(n_i)\}_{i=0}^{\infty}$ is a descending chain in B; but it is also recursive, since

$$b(n_0) = b_0,$$
$$b(n_{i+1}) = b_r,$$

where

$$r = \mu_s\{s > t. \& .u < s \Rightarrow b_u \neq b_s\} \text{ and } b_t = b(n_i).$$

This is impossible since B is a quasi-well-ordering.

The second assertion in the theorem is clear from definition 6.1.1.

6.1.3 THEOREM. There is a partial recursive functional Ψ such that if $p:\mathsf{A}_0 \simeq \mathsf{A}_1$ and $q:\mathsf{B}_0 \simeq \mathsf{B}_1$ then

$$\Psi(p, q): \mathsf{A}_0\cdot\mathsf{B}_0 \simeq \mathsf{A}_1\cdot\mathsf{B}_1.$$

PROOF. Set $\Psi(p, q)(x) = j(pk(x), ql(x))$.

Theorem 6.1.3. permits the following definition.

6.1.4 DEFINITION. $A\cdot B = \mathrm{COT}(\mathsf{A}\cdot\mathsf{B})$ for any $\mathsf{A} \in A$ and $\mathsf{B} \in B$.

We shall often write AB for $A\cdot B$ and AB for $\mathsf{A}\cdot\mathsf{B}$.

6.1.5 THEOREM. If $\mathsf{A}, \mathsf{B} \subseteq \mathsf{R}$, then

$$\mathsf{A}\cdot\mathsf{B} = \mathsf{R}\cdot\mathsf{R} \cap j(\mathrm{A}, \mathrm{B})^2,$$

where $\mathrm{A} = \mathrm{C}\text{‘}\mathsf{A}$, $\mathrm{B} = \mathrm{C}\text{‘}\mathsf{B}$ and

$$j(\mathrm{A}, \mathrm{B}) = \{j(a, b): a \in \mathrm{A} \ \&\ b \in \mathrm{B}\}.$$

PROOF. Left to the reader.

6.1.6 THEOREM. $\mathsf{R}\cdot\mathsf{R} \simeq \mathsf{R}$.

PROOF. $\mathsf{R}\cdot\mathsf{R}$ is a recursive linear ordering by theorem 6.1.2. Further, if

$$x <_{\mathsf{R}\cdot\mathsf{R}} y \quad \text{and} \quad x = j(a, b), \quad y = j(a', b'),$$

then either $b \prec b'$ when, since R is dense, there is a c such that

$$b \prec c \prec b',$$

or $b=b'$ and $a \prec a'$ and there is a d such that

$$a \prec d \prec a'.$$

Hence either

$$x <_{\mathsf{R}\cdot\mathsf{R}} j(a, c) <_{\mathsf{R}\cdot\mathsf{R}} y,$$

or

$$x <_{\mathsf{R}\cdot\mathsf{R}} j(d, b) <_{\mathsf{R}\cdot\mathsf{R}} y$$

and so $\mathsf{R}\cdot\mathsf{R}$ is dense. Clearly $j(0, 0)$ is the first element of $\mathsf{R}\cdot\mathsf{R}$ and if $j(a, b) \in C`\mathsf{R}\cdot\mathsf{R}$ and

$$b = \langle b(0), \ldots, b(m)\rangle,$$

then

$$j(a, b) <_{\mathsf{R}\cdot\mathsf{R}} j(a, \langle b(0), \ldots, b(m) + 1\rangle),$$

so $\mathsf{R}\cdot\mathsf{R}$ has no last element. By theorem 1.2.4 it follows that $\mathsf{R}\cdot\mathsf{R} \simeq \mathsf{R}$.

Since $\mathsf{R}\cdot\mathsf{R}$ and R are recursive linear orderings there is a minimal[61] recursive isomorphism (i.e. with domain $C`\mathsf{R}\cdot\mathsf{R}$ and range $C`\mathsf{R}$)

$$m: \mathsf{R}\cdot\mathsf{R} \simeq \mathsf{R}.$$

We shall regard this m as fixed from now on. Now suppose we are given $\mathsf{A}, \mathsf{B} \subseteq \mathsf{R}$, then by theorem 6.1.6 there exists a C such that

$$m: \mathsf{A}\cdot\mathsf{B} \simeq \mathsf{C} \subseteq \mathsf{R},$$

where $C`\mathsf{C} = m(j(C`\mathsf{A}, C`\mathsf{B}))$. We denote this (unique) C by $m(\mathsf{A}, \mathsf{B})$ whenever $\mathsf{A}, \mathsf{B} \subseteq \mathsf{R}$. Clearly, by theorem 6.1.3, if $m': \mathsf{R}\cdot\mathsf{R} \simeq \mathsf{R}$ and $m'(\mathsf{A}, \mathsf{B})$ is defined analogously to $m(\mathsf{A}, \mathsf{B})$ then for some recursive isomorphism p

$$p: m(\mathsf{A}, \mathsf{B}) \simeq m'(\mathsf{A}, \mathsf{B}),$$

where p may be taken to have range and domain Seq and to be independent of A, B, in fact take

$$p = m'm^{-1}.$$

6.1.7 LEMMA. $(\mathsf{A}_0\cdot\mathsf{A}_1)\cdot\mathsf{A}_2 \simeq \mathsf{A}_0\cdot(\mathsf{A}_1\cdot\mathsf{A}_2)$.

PROOF. $\langle x, y\rangle \in (\mathsf{A}_0\cdot\mathsf{A}_1)\cdot\mathsf{A}_2$

$\Leftrightarrow x = j(j(a_0, a_1), a_2) \;\&\; y = j(j(a_0', a_1'), a_2')$

$\&\; a_i, a_i' \in C`\mathsf{A}_i \quad (i = 0, 1, 2) \;\&\; [a_2 <_{\mathsf{A}_2} a_2'$

$\vee (a_2 = a_2' \;\&\; a_1 <_{\mathsf{A}_1} a_1') \vee (a_2 = a_2' \;\&\; a_1 = a_1' \;\&\; a_0 \leq_{\mathsf{A}_0} a_0')]$.

$$\begin{aligned}&\langle x, y\rangle \in \mathsf{A}_0\cdot(\mathsf{A}_1\cdot\mathsf{A}_2)\\ &\Leftrightarrow x = j(a_0, j(a_1, a_2))\ \&\ y = j(a', j(a_1', a_2'))\\ &\&\ a_i, a_i' \in \mathrm{C}‘\mathsf{A}_i \quad (i = 0, 1, 2)\\ &\&\ [a_2 <_{\mathsf{A}_2} a_2' \vee (a_2 = a_2'\ \&\ a_1 <_{\mathsf{A}_1} a_1')\\ &\vee (j(a_1, a_2) = j(a_1', a_2')\ \&\ a_0 \leq_{\mathsf{A}_0} a_0')].\end{aligned}$$

Since j is one-one, the conditions in square brackets are equivalent and the proof is completed by using the recursive isomorphism

$$x \to j(kk(x), j(lk(x), l(x))).$$

6.1.8 THEOREM. Multiplication is associative, i.e.

$$(A\cdot B)\cdot C = A\cdot(B\cdot C).$$

PROOF. Immediate from theorem 6.1.3 and lemma 6.1.7. Multiplication is not commutative since we know classically $\omega\cdot 2 \neq 2\cdot\omega = \omega$. This theorem allows us to omit brackets in writing products of several C.O.T.s.

6.1.9 THEOREM. $A\cdot B = 0 \Leftrightarrow A = 0 \vee B = 0$.

PROOF. $j(\mathrm{A}, \mathrm{B}) = \emptyset \Leftrightarrow \mathrm{A} = \emptyset \vee \mathrm{B} = \emptyset$.

6.2.

6.2.1 THEOREM. $A(B+C) = AB + AC$.

PROOF. Take $\mathsf{A}, \mathsf{B}, \mathsf{C} \subseteq \mathsf{R}$, then

$$p: \mathsf{A}\cdot(\mathsf{B} + \mathsf{C}) \simeq \mathsf{A}\cdot\mathsf{B} + \mathsf{A}\cdot\mathsf{C},$$

where p is defined as follows

$$\begin{aligned} p(x) &= \hat{0}(j(y, z)) \quad \text{if} \quad x = j(y, \hat{0}(z)),\\ &= \hat{1}(j(y, z)) \quad \text{if} \quad x = j(y, \hat{1}(z)),\\ &\text{is undefined otherwise.}\end{aligned}$$

We leave the reader to verify that p is one-one; the other properties are obvious.

The other distributive law does not hold since classically we have $(\omega + 1)\,\omega = \omega^2 \neq \omega^2 + \omega$.

6.2.2 THEOREM. (i) $A = A \cdot 1$,
(ii) $A \cdot n + A = A + A \cdot n = A \cdot (n+1)$,
(iii) $A \cdot m + A \cdot n = A \cdot (m+n)$,
(iv) $A \cdot W = A + A \cdot W = A \cdot n + A \cdot W$ for all n,
(v) $A \cdot (mn) = (A \cdot m) \cdot n = (A \cdot n) \cdot m$,
(vi) $A \cdot W = (A \cdot n) \cdot W$ for all n.

PROOF. (i), (ii) Left to the reader.
(iii) By theorem 6.2.1.
(iv) By (i) and theorem 6.2.1 we have

$$A \cdot 1 + A \cdot W = A \cdot (1 + W) = A \cdot W$$

by theorem 5.2.6. The other part follows by induction on n and is left to the reader.
(v) By theorem 6.1.8.
(vi) Let $\mathsf{A} \in A$ and let d, r be the (primitive recursive) functions such that

$$x = nd(x) + r(x) \quad \text{and} \quad 0 \le r(x) < n$$

and let

$$p(x) = j(j(k(x), rl(x)), dl(x)).$$

then p is one-one and (primitive) recursive. We assert that p is order preserving between $\mathsf{A} \cdot \mathsf{W}$ and $(\mathsf{A} \cdot \mathsf{I}_n) \cdot \mathsf{W}$ where

$$\mathsf{I}_n = \{\langle i, j \rangle : i \le j < n\}$$

and

$$\mathsf{W} = \{\langle i, j \rangle : i \le j\},$$

for

$$\begin{aligned}(\mathsf{A} \cdot \mathsf{I}_n) \cdot \mathsf{W} = \{\langle j(j(a, s), u), j(j(b, t), v) \rangle : s, t < n \,.\&.\\ (u < v \vee u = v \;\&\; s < t) \;\&\; a, b \in \mathrm{C}`\mathsf{A} \\ .\vee.\, u = v \;\&\; s = t \;\&\; \langle a, b \rangle \in \mathsf{A}\}\end{aligned} \tag{1}$$

and

$$\langle x, y \rangle \in \mathsf{A} \cdot \mathsf{W} \Leftrightarrow x = j(a, nq + r) \;\&\; y = j(a', nq' + r'),$$

where

$$\begin{aligned}0 \le r, r' < n \;\&\; nq + r < nq' + r' \;\&\; a, a' \in \mathrm{C}`\mathsf{A} \\ .\vee.\, nq + r = nq' + r' \;\&\; \langle a, a' \rangle \in \mathsf{A}.\end{aligned} \tag{2}$$

Condition (2) is equivalent to

$$\begin{aligned}0 \le r, r' < n \,.\&.\, (q < q' \vee q = q' \;\&\; r < r') \;\&\; a, a' \in \mathrm{C}`\mathsf{A} \\ .\vee.\, q = q' \;\&\; r = r' \;\&\; \langle a, a' \rangle \in \mathsf{A}.\end{aligned} \tag{3}$$

Comparison of (1) and (3) immediately shows that p is order preserving. This completes the proof.

6.2.3 THEOREM. (i) $(A+B)\cdot n+A=A+(B+A)\cdot n$,
(ii) $(A+B)\cdot(n+1)=A+(B+A)\cdot n+B$,
(iii) $(A+B)\cdot W=A+(B+A)\cdot W$.

PROOF. (i), (ii) Proof by induction using the associativity of addition.

(iii) Let $\mathsf{A}\in A$, $\mathsf{B}\in B$ where $\mathsf{A}, \mathsf{B}\subseteq \mathsf{R}$, then

$$p:(\mathsf{A}+\mathsf{B})\cdot\mathsf{W}' \simeq \mathsf{A}+(\mathsf{B}+\mathsf{A})\cdot\mathsf{W}',$$

where

$$\begin{aligned} p(x) &= \hat{0}(y) && \text{if } x=j(\hat{0}(y), \langle 0\rangle),\\ &= j(\hat{1}\hat{0}(y), \langle z\rangle) && \text{if } x=j(\hat{1}(y), \langle z\rangle),\\ &= j(\hat{1}\hat{1}(y), \langle z \dot{-} 1\rangle) && \text{if } x=j(\hat{0}(y), \langle z\rangle)\ \&\ z\geq 1, \end{aligned}$$

is undefined otherwise.

We leave the reader to check that p has the required properties.

6.2.4 THEOREM. If $A+B=B+A$, then
(i) $A\cdot m+B\cdot n=B\cdot n+A\cdot m$,
(ii) $(A+B)\cdot m=A\cdot m+B\cdot m$.

PROOF. By induction using theorem 6.2.2.(ii).

6.2.5 THEOREM. If $B\cdot W=A+C$ and $C\neq 0$, then there exist n, D, E such that

$$A=B\cdot n+D, \quad D+E=B \quad \text{and} \quad E+B\cdot W=C.$$

PROOF. We consider only the non-trivial case when A, B, C are all non-zero. Let $\mathsf{B}\in B$, then by the separation lemma (2.3.1) there exist $\mathsf{A}\in A$ and $\mathsf{C}\in C$ such that

$$\mathsf{A})(\mathsf{C} \quad \text{and} \quad \mathsf{B}\cdot\mathsf{W}=\mathsf{A}\mathbin{\hat{+}}\mathsf{C},$$

where W is as in the proof of theorem 6.2.2.(vi). By assumption

$$\mathsf{C}\neq\emptyset\neq\mathsf{A},$$

hence there is a $c\in C$‘C where $c=j(b, n')$ for some $b\in C$‘B and some $n'\geq 0$. Therefore

$$\mathrm{F}=\{l(a): a\in C‘\mathsf{A}\}$$

is a set of natural numbers bounded by n'. F therefore has a maximum, let it be n. Further let

$$\mathsf{D} = \mathsf{A}[j(\mathrm{B}, n) \quad \text{and} \quad \mathsf{E} = \mathsf{C}[j(\mathrm{B}, n),$$

where $\mathrm{B} = \mathrm{C}`\mathsf{B}$ and

$$j(\mathrm{B}, n) = \{j(b, n) : b \in \mathrm{B}\}.$$

Then, as in the proof of the directed refinement theorem (2.3.2) it is easily verified that

$$\mathsf{D})(\mathsf{E},\ \mathsf{A} = \mathsf{B}\cdot n \mathbin{\hat{+}} \mathsf{D},\ \mathsf{D} \mathbin{\hat{+}} \mathsf{E} = \mathsf{B} \quad \text{and} \quad \mathsf{E} \mathbin{\hat{+}} \mathsf{B}\cdot\mathsf{W}[\{x : l(x) > n\} = \mathsf{C}. \qquad (1)$$

We observe that

$$\mathsf{B}\cdot\mathsf{W}[\{x : l(x) > n\} \simeq \mathsf{B}\cdot\mathsf{W}$$

under the map

$$p : x \to j(k(x), l(x) \dot{-} (n+1))$$

defined only on

$$\{x : l(x) > n\}.$$

Taking C.O.T.s of both sides of the equations in (1) completes the proof.

The above theorem is not an immediate consequence of the directed refinement theorem for the following reason. We know that

$$B\cdot n + B\cdot W = B\cdot W \quad \text{for all} \quad n.$$

Suppose $B\cdot W = A + C$ with $C \neq 0$, then it follows easily from the directed refinement theorem (2.3.2) that for each n either

$$A \leq B\cdot n \quad \text{or} \quad B\cdot n \leq A.$$

If $A \leq B\cdot n$ for some n, then we are through, but $A \geq B\cdot n$ for all n does not imply $A \geq B\cdot W$ for take $A = V$, $B = 1$.

6.2.6 Corollary. If A is a co-ordinal and $A < B\cdot W$ then $A \leq B\cdot n$ for some n.

6.2.7 Theorem. If C is a recursive quord or a co-ordinal, then

$$C \leq^* B\cdot W \;\&\; C \neq 0 \Rightarrow B\cdot W \leq^* C.$$

PROOF. By theorem 4.2.1,

$$C \leq^* B\cdot W \Rightarrow (\exists A)\,(A + C = B\cdot W).$$

Hence by theorem 6.2.5 there is an E such that

$$E + B\cdot W = C, \quad \text{hence} \quad B\cdot W \leq^* C.$$

6.2.8 COROLLARY. If $A \neq 0$, $A \leq B\cdot W$ and $A \leq^* B\cdot W$, then $A = B\cdot W$.
PROOF. Immediate from theorem 4.2.6 and the theorem above.

We now complete our demonstration that $\leq^*$ is not antisymmetric. V and W are incomparable but

$$V < (V + W)\cdot W \quad \text{and} \quad W < W + (V + W)\cdot W,$$

hence since $\leq$ is a tree ordering

$$(V + W)\cdot W \neq W + (V + W)\cdot W. \tag{2}$$

Now

$$V + W + (V + W)\cdot W = (V + W)\cdot W,$$

hence

$$W + (V + W)\cdot W \leq^* (V + W)\cdot W.$$

But

$$(V + W)\cdot W \leq^* W + (V + W)\cdot W,$$

hence if $\leq^*$ is antisymmetric (2) is false, which is impossible.[62]

6.3 We now present a lemma which we shall use in giving existence proofs later on.

6.3.1 LEMMA. If A is a quord, then

$$B + A = A \Leftrightarrow (\exists C)\,(B\cdot W + C = A).$$

If A is a recursive quord then

$$B + A = A \Leftrightarrow B\cdot W \leq A.$$

PROOF. The second statement follows immediately from the first. Suppose $B\cdot W + C = A$, then

$$B + A = B + (B\cdot W + C) = (B + B\cdot W) + C = B\cdot W + C = A.$$

Now suppose $B+A=A$. If $B=0$, then the assertion is trivial. Similarly if $A=0$, so we assume neither of these occurs. By hypothesis there exist quasi-well-orderings A, $\mathsf{B}\subseteq\mathsf{R}$ and a recursive isomorphism p such that

$$p:\mathsf{B}\mathbin{\hat{+}}\mathsf{A}\simeq\mathsf{A}$$

where $\mathsf{A}\in A$ and $\mathsf{B}\in B$ and $\mathsf{A})\,(\mathsf{B}$. Let $\mathrm{A}=\mathrm{C}`\mathsf{A}$ and $\mathrm{B}=\mathrm{C}`\mathsf{B}$.

We introduce the following notation for this proof only:

$$\begin{aligned}
\mathrm{B}_\omega &= \bigcup_{n=0}^{\infty} p^{n+1}(\mathrm{B}),\\
\mathsf{B}_\omega &= \mathsf{A}[\mathrm{B}_\omega,\\
\mathrm{A}_0 &= \{x:(\forall n)\,(p^{-n}(x)\in\mathrm{A})\},\\
\mathsf{A}_0 &= \mathsf{A}[\mathrm{A}_0.
\end{aligned}$$

Before giving the formal details we sketch the idea of the proof. Classically, if α, β are ordinals then

$$\beta+\alpha=\alpha\Rightarrow\beta\cdot n<\alpha\quad\text{for all}\quad n.$$

Hence

$$\lim\beta\cdot n\leq\alpha$$

so

$$\beta\cdot\omega+\gamma=\alpha\quad\text{for some}\quad\gamma.$$

Now, firstly (as we shall show in chapter 12) in general we cannot uniquely define limits (though in this case we could) and secondly even if a limit had been defined then we should still have to establish separability of the orderings representing $\beta\cdot\omega$ and γ. In this particular case, however, we can establish both fairly easily by an *ad hoc* device. We observe that

$$\beta+(\beta\cdot\omega+\gamma)=\beta\cdot\omega+\gamma$$

and in the case of any particular ordering of type α the elements in that ordering belonging to the final segment of type γ all remain fixed under any map p such that

$$p:\mathsf{B}\mathbin{\hat{+}}\mathsf{A}\simeq\mathsf{A}.$$

So in order to separate the $\beta\cdot\omega$ and γ orderings we simply look at the sets of elements which remain fixed under p and those which do not. These two sets include A_0 and B_ω, respectively. To effect the proof we shall establish

1) $\mathrm{A}_0\cap\mathrm{B}_\omega=\emptyset$,
2) $\mathrm{A}_0\cup\mathrm{B}_\omega=\mathrm{A}$,

3) $\mathsf{B}_\omega \in B \cdot W$,
4) $x \in \mathrm{A}_0 \Rightarrow p(x) = x$,
5) $x \in \mathrm{B} \Rightarrow p(x) \neq x$,
6) $\mathsf{B}_\omega)(\mathsf{A}_0$,
7) $\mathsf{B}_\omega \dotplus \mathsf{A}_0 = \mathsf{A}$.

1) If $x \in \mathrm{B}$, then $x = p^n(y)$ for some $n > 0$, some $y \in \mathrm{B}$. Hence $p^{-n}(x)$ is defined and $\notin \mathrm{A}$; so $x \notin \mathrm{A}_0$.

2) Since p maps $\mathrm{B} \cup \mathrm{A}$ *onto* A, $x \in \mathrm{A}$ implies either

$$(\forall n)\,[p^{-n}(x) \in \mathrm{A}],$$

or

$$(\exists n)\,[p^{-n}(x) \in \mathrm{B}].$$

I.e. $x \in \mathrm{A} \Rightarrow x \in \mathrm{A}_0 \vee x \in \mathrm{B}$.

Conversely,

$$x \in \mathrm{A}_0 \Rightarrow x = p^0(x) \in \mathrm{A}$$

and

$$x \in \mathrm{B}_\omega \Rightarrow x = p^n(y) \quad \text{for some} \quad y \in \mathrm{B}, \quad \text{some} \quad n > 0,$$

so in either case $x \in \mathrm{A}$.

3) Since $\mathsf{B})(\mathsf{A}$ there is a partial recursive function f such that if $x \in \mathrm{B} \cup \mathrm{A}$ then

$$x \in \mathrm{A} \Leftrightarrow f(x) = 0 \;\&\; x \in \mathrm{B} \Leftrightarrow f(x) = 1$$

by theorem 2.1.3. We now use f to calculate a function q such that

$$q : \mathsf{B}_\omega \simeq \mathsf{B} \cdot \mathsf{W}.$$

Let

$$q(x) = j(p^{-n}(x), n-1),$$

where

$$n = \mu_r \{ r \geq 1 \;\&\; f(p^{-r}(x)) = 0 \;\&\; (\forall s)\,(s < r \Rightarrow f(p^{-s}(x)) = 1) \}.$$

(n is undefined if *any* conjunct is undefined.)
Clearly, q is partial recursive. Suppose $q(x) = q(y)$, then

$$q(x) = j(x', n) = q(y) \quad \text{for some} \quad x' = p^{-n-1}(x) = p^{-n-1}(y).$$

But p^{-1} is one-one, therefore $x = y$ and q is one-one.

We now show q maps B_ω onto $\mathrm{C}‘\mathsf{B} \cdot \mathsf{W}$. Suppose $x \in \mathrm{B}_\omega$; then both conjuncts in the definition of n, above, are defined and hence $q(x)$ is defined. Clearly, $q(x) \in \mathrm{C}‘\mathsf{B} \cdot \mathsf{W}$. Now suppose $j(x, n) \in \mathrm{C}‘\mathsf{B} \cdot \mathsf{W}$; then $p^{n+1}(x) \in \mathrm{B}_\omega$ and $q(p^{n+1}(x)) = j(x, n)$; hence $\mathrm{C}‘\mathsf{B} \cdot \mathsf{W} \subseteq q(\mathrm{B}_\omega)$.

Next we show that q is order-preserving between B_ω and $\mathsf{B}\cdot\mathsf{W}$. It suffices to show that if

$$\langle x_0, y_0\rangle \in \mathsf{B}_\omega \ \&\ x_0 = p^m(x) \ \&\ y_0 = p^n(y),$$

where $x, y \in \mathrm{B}$ and

$$0 < r < m \Rightarrow p^r(x) \in \mathrm{A} \quad \text{and} \quad 0 < s < n \Rightarrow p^s(y) \in \mathrm{A},$$

then

$$1 \le m < n \quad \text{or} \quad 1 \le m = n \ \&\ \langle x, y\rangle \in \mathsf{B}.$$

If $m > n$, then since p is one-one and order-preserving,

$$\langle p^{m-n}(x), y\rangle \in \mathsf{B} \mathbin{\hat{+}} \mathsf{A}.$$

But

$$y \in \mathrm{B} \quad \text{and} \quad p^{m-n}(x) \in \mathrm{A},$$

therefore, by the antisymmetry of $\mathsf{B} \mathbin{\hat{+}} \mathsf{A}$,

$$y = p^{m-n}(x),$$

which contradicts $\mathsf{B})(\mathsf{A}$. Thus $m \le n$. If $m = n$, then

$$\langle x, y\rangle \in \mathsf{B} \mathbin{\hat{+}} \mathsf{A} \quad \text{where} \quad x, y \in \mathrm{B}.$$

We conclude $\langle x, y\rangle \in \mathsf{B}$. This completes the proof of 3).

4) Since A is a quasi-well-ordering and $\mathsf{A}_0 \subseteq \mathsf{A}$, A_0 is also a quasi-well-ordering. Now p maps $\mathrm{A}_0 = \mathrm{C}‘\mathsf{A}_0$ onto A_0 since

$$x \in \mathrm{A}_0 \Rightarrow p^{-1}(x) \in \mathrm{A}_0 \ \&\ p(x) \in \mathrm{A}_0$$

which implies

$$\mathrm{A}_0 \subseteq p(\mathrm{A}_0) \subseteq \mathrm{A}_0.$$

Now p is order preserving, hence by lemma 3.2.6 p, p^{-1} are both increasing on A_0 so we conclude p is the identity on A_0.

5) $x \in \mathrm{B} \Rightarrow x = p^n(y)$ for some $n > 0$, some $y \in \mathrm{B}$. Since p is one-one

$$x = p(x) \quad \text{implies} \quad p^{-n}(x) = p^{-n+1}(x).$$

But

$$p^{-n}(x) \in \mathrm{B},\ p^{-n+1}(x) \in \mathrm{A} \quad \text{and} \quad \mathrm{B} \cap \mathrm{A} = \emptyset \quad \text{since} \quad \mathsf{B})(\mathsf{A}.$$

We conclude $p(x) \neq x$.

6) Since p is partial recursive, δp is r.e. Hence by 4) and 5)

$$\mathrm{A}_0 \subseteq \{x : x \in \delta p \ \&\ p(x) \neq x\}$$

and

$$\mathsf{B}_\omega \subseteq \{x : x \in \delta p \ \&\ p(x) = x\}$$

and the sets on the right hand sides are r.e. Thus $\mathsf{B}_\omega)\ (\mathsf{A}_0$.

7) By 6), $\mathsf{B}_\omega \dotplus \mathsf{A}_0$ is well-defined. By 2) $C`(\mathsf{B}_\omega \dotplus \mathsf{A}_0) = \mathsf{A}$. By definition $\mathsf{B}_\omega \subseteq \mathsf{A}$ and $\mathsf{A}_0 \subseteq \mathsf{A}$. It therefore suffices to prove that

$$\mathsf{B}_\omega \times \mathsf{A}_0 \subseteq \mathsf{A} \quad \text{and} \quad \mathsf{A} \subseteq \mathsf{B}_\omega \dotplus \mathsf{A}_0 .$$

If $x \in \mathrm{B}_\omega$ and $y \in \mathrm{A}_0$ then

$$(\exists n)\,[p^{-n}(x) \in \mathrm{B}] \quad \text{and} \quad (\forall n)\,[p^{-n}(y) \in \mathrm{A}] .$$

Hence

$$\langle p^{-n}(x), p^{-n}(y)\rangle \in \mathrm{B} \times \mathrm{A} \subseteq \mathsf{B} \dotplus \mathsf{A} \quad \text{for some} \quad n ,$$

and since p is order preserving we have $\langle x, y\rangle \in \mathsf{A}$. If $\langle x, y\rangle \in \mathsf{A}$, then either (i) $x, y \in \mathrm{B}_\omega$ or (ii) $x \in \mathrm{B}_\omega$, $y \in \mathrm{A}_0$ or (iii) $x, y \in \mathrm{A}_0$ or (iv) $x \in \mathrm{A}_0, y \in \mathrm{B}_\omega$ since $\mathrm{A} = \mathrm{A}_0 \cup \mathrm{B}_\omega$ by 2). Hence in order to complete the proof of 7) we only need to show (iv) is impossible. If (iv) holds, then there is an n such that

$$p^{-n}(x) \in \mathrm{A} \quad \text{and} \quad p^{-n}(y) \in \mathrm{B} ,$$

which is impossible since p is order preserving and

$$(\mathsf{A} \times \mathsf{B}) \cap (\mathsf{B} \dotplus \mathsf{A}) = \emptyset .$$

We now complete the proof of the lemma. By 3) $\mathsf{B}_\omega \in B \cdot W$. Let

$$C = \mathrm{COT}(\mathsf{A}_0), \quad \text{then by 7)}, \quad B \cdot W + C = A .$$

6.3.2 THEOREM. If C is a quord, $A + B + C = C$ and $(A + B) \cdot W = (B + A) \cdot W$, then

$$A + C = B + C = C .$$

PROOF.[63] By lemma 6.3.1 and the first part of the hypothesis there is a D such that

$$(A + B) \cdot W + D = C .$$

Hence by the second part of the hypothesis, with the same D,

$$(B + A) \cdot W + D = C . \tag{1}$$

Now by theorem 6.2.3.(iii),

$$(A + B) \cdot W = A + (B + A) \cdot W ,$$

hence

$$A + (B + A)\cdot W + D = (A + B)\cdot W + D = C,$$

and therefore

$$A + C = C.$$

Similarly, from (1) we conclude

$$B + C = C.$$

6.3.3 THEOREM. If $A\cdot m+B\leq B$ (or equivalently, $A\cdot m+B=B$) and $m>0$, then $A\cdot n+B=B$ for all n.

PROOF. Clearly, $B\leq^* A\cdot m+B$, hence by theorem 4.2.6, $B=A\cdot m+B$. From lemma 6.3.1 we obtain

$$(\exists C)\,[(A\cdot m)\cdot W + C = B],$$

but by theorem 6.2.2.(vi),

$$(A\cdot m)\cdot W = A\cdot W$$

and therefore by lemma 6.3.1,

$$A + B = B.$$

An easy induction now establishes

$$A\cdot n + B = B \quad \text{for all} \quad n.$$

6.3.4 THEOREM. If $A\cdot W+C=B\cdot m$ for some m, where B is a co-ordinal, then $A\cdot W+C'=B$ for some C'.

PROOF (by induction on m). If $m=0$ or 1 then the assertion is trivial. Now suppose $m>0$ and the assertion holds for m. Then

$$A\cdot W + C = B\cdot m + B.$$

Hence by the directed refinement theorem (2.3.2) there is an E such that either

$$A\cdot W + E = B\cdot m,$$

or

$$A\cdot W = B\cdot m + E \;\&\; E + C = B \;\&\; E \neq 0.$$

In the former case we are through by the induction hypothesis; in the latter

$$E \leq^* A\cdot W$$

and hence by theorem 6.2.5, since $B \cdot m$ is a co-ordinal,

$$F + A \cdot W = E \quad \text{for some (co-ordinal) } F.$$

Now $E + C = B$ implies

$$F + A \cdot W + C = B,$$

hence

$$A \cdot W + C \leq^* B.$$

However,

$$B \leq B \cdot m \leq A \cdot W \leq A \cdot W + C$$

and hence by theorem 4.2.6,

$$A \cdot W + C = B.$$

CHAPTER 7

EXPONENTIATION

7.1 In this chapter we define exponentiation and give PARIKH's proof (1962, 1966) that the collection of (recursive) quords is not closed under exponentiation. The classical method of defining exponentiation of orderings is by means of an ordering of finite sequences of elements from the given orderings. Since we are dealing only with quasi-well-orderings which can be embedded in R by recursive isomorphisms we give a definition in the spirit of theorem 6.1.5.

Convention. We shall assume from now on that if $\mathsf{A} \subseteq \mathsf{R}$ and A has a first element then this first element is 0, unless otherwise stated.

We write $\min(\mathsf{B})$ for the first element of an arbitrary ordering B if it exists. As in the proof of theorem 1.2.4 we can always assume that if

$$p: \mathsf{A} \simeq \mathsf{A}' \subseteq \mathsf{R} \quad \text{then} \quad p(\min(\mathsf{A})) = 0.$$

We shall also need to code finite sequences of ordered pairs of numbers so we introduce *bracket symbols.*

7.1.1 DEFINITION. A symbol of the form

$$\begin{pmatrix} b_0 \dots b_n \\ a_0 \dots a_n \end{pmatrix}, \tag{1}$$

where $n \geq -1$ and the a_i, b_i $(i=0, \dots, n)$ are natural numbers is said to be a *bracket symbol.* If the symbol is simply ( ) (i.e. $n=-1$) then we call this the *empty* bracket symbol and denote it by $\mathbf{0}$. We use upper case bold face letters (*A*, *B*, *C*,...) for bracket symbols. We write

$$(\mathsf{A}, \mathsf{B})$$

for the collection of all bracket symbols (including the empty one) of the form (1) where

$$a_i \in C`\mathsf{A}, \quad b_i \in C`\mathsf{B}, \quad a_i \neq \min(\mathsf{A}) \quad i = 0, \ldots, n,$$
$$b_i \neq b_{i+1} \quad \text{and} \quad \langle b_{i+1}, b_i \rangle \in \mathsf{B}, \quad i = 0, \ldots, n-1.$$

We define our coding by

$$e(0) = 0,$$

and if

$$\boldsymbol{A} = \begin{pmatrix} b_0 \ldots b_n \\ a_0 \ldots a_n \end{pmatrix}, \quad a_i \neq 0,$$
$$e(\boldsymbol{A}) = \langle j(a_0, b_0), \ldots, j(a_n, b_n) \rangle$$
$$= \prod_{i=0}^{n} p_i^{j(a_i, b_i)},$$

where p_i denotes the ith prime ($p_0 = 2$).[71]

7.1.2 THEOREM. *e is a one-one primitive recursive function from the set of all bracket symbols with the $a_i \neq 0$ into $\mathscr{N}$. Further, ρe is recursive.*

PROOF. The a_i, b_i and n are uniquely recoverable from any element x in ρe since if $x \neq 0$ then

$$a_i = k((x)_i), \, b_i = l((x)_i) \quad \text{and} \quad n = lh(x) + 1.$$

The other parts of the assertion are left to the reader to prove.

We shall write

$$b_0 \succ b_1 \cdots \succ b_n$$

for

$$b_0 \succ b_1 \,\&\, \cdots \,\&\, b_{n-1} \succ b_n$$

and similarly write

$$b_0 >_{\mathsf{B}} b_1 \cdots >_{\mathsf{B}} b_n$$

in the obvious way.

7.1.3 DEFINITION.

$$\mathsf{R}^{\mathsf{R}} = \mathsf{R} \exp \mathsf{R} =$$
$$= \{\langle e(\boldsymbol{K}), e(\boldsymbol{K}') \rangle : \boldsymbol{K} = \begin{pmatrix} b_0 \ldots b_m \\ a_0 \ldots a_m \end{pmatrix} \,\&\, \boldsymbol{K}' = \begin{pmatrix} b'_0 \ldots b'_n \\ a'_0 \ldots a'_n \end{pmatrix}$$

$$\& \boldsymbol{K}, \boldsymbol{K}' \in (\mathsf{R}, \mathsf{R}) \; \& \; \{\boldsymbol{K} = \mathbf{0} \vee [b_0 \succ b_1 \succ \cdots \succ b_m \; \& \; b'_0 \succ b'_1 \succ \cdots \succ b'_n$$
$$\& [\boldsymbol{K} \neq \mathbf{0} \; \& \; \{(m \leq n \; \& \; (\forall r)(r \leq m \Rightarrow a_r = a'_r \; \& \; b_r = b'_r))$$
$$\vee (\exists r)(\forall t)[(t < r \Leftrightarrow (\forall s \leq t)(a_s = a'_s \; \& \; b_s = b'_s))$$
$$\& (b_r \prec b'_r \vee [b_r = b'_r \; \& \; a_r \preccurlyeq a'_r])]\}]\}\}.$$

If $\mathsf{A}, \mathsf{B} \subseteq \mathsf{R}$ and $\min(\mathsf{A}) = 0$

$$\mathsf{A}^{\mathsf{B}} = \mathsf{A} \exp \mathsf{B} = \mathsf{R} \exp \mathsf{R} \cap (\mathsf{A}, \mathsf{B})^2 .$$

7.1.4 THEOREM. There is a partial recursive functional Ξ such that if $p: \mathsf{A}_0 \simeq \mathsf{A}_1$ and $q: \mathsf{B}_0 \simeq \mathsf{B}_1$, A_i, $\mathsf{B}_i \subseteq \mathsf{R}$, and $\min(\mathsf{A}_i) = 0$ $(i = 0, 1)$. Then

$$\Xi(p, q) : \mathsf{A}_0 \exp \mathsf{B}_0 \simeq \mathsf{A}_1 \exp \mathsf{B}_1 .$$

PROOF. Let

$$\Xi(p, q)(0) = 0,$$

$$\Xi(p, q)(x) = e\begin{pmatrix} q(b_0) \dots q(b_n) \\ p(a_0) \dots p(a_n) \end{pmatrix} \quad \text{if} \quad x \in \rho e, \; x = e\begin{pmatrix} b_0 \dots b_n \\ a_0 \dots a_n \end{pmatrix},$$

is undefined otherwise.

That Ξ is well-defined follows at once from theorem 7.1.2 and the rest of the proof is left to the reader.

7.1.5 DEFINITION. If $p: \mathsf{A}_0 \simeq \mathsf{A}_1$ and $q: \mathsf{B}_0 \simeq \mathsf{B}_1$, where A_0, $\mathsf{B}_0 \subseteq \mathsf{R}$, and $\min(\mathsf{A}_i) = 0$ $(i = 0, 1)$ then we set

$$\mathsf{A}_1 \exp \mathsf{B}_1 = \Xi(p, q)(\mathsf{A}_0 \exp \mathsf{B}_0).$$

This definition does not define $\mathsf{A}_1 \exp \mathsf{B}_1$ uniquely, though we could easily arrange this by requiring, e.g. that p, q be recursive isomorphisms with least Gödel numbers; however, this is unimportant as up to recursive isomorphism the definition is unique as the next theorem shows.

7.1.6 THEOREM. If $p: \mathsf{A}_0 \simeq \mathsf{A}_1$ and $q: \mathsf{B}_0 \simeq \mathsf{B}_1$ where A_i, B_i are arbitrary quasi-well-orderings (embeddable in R by some recursive isomorphisms) and $\min(\mathsf{A}_i) = 0$ $(i = 0, 1)$ then

$$\mathsf{A}_0 \exp \mathsf{B}_0 \simeq \mathsf{A}_1 \exp \mathsf{B}_1$$

PROOF. Left to the reader.

7.2

7.2.1 THEOREM. (i) If A, B are well-orderings then $\mathsf{A} \exp \mathsf{B}$ is a well-ordering.

(ii) (PARIKH, 1962, 1966). If B is a well-ordering and A is a quasi-well-ordering, then $\mathsf{A} \exp \mathsf{B}$ is a quasi-well-ordering.

(iii) (PARIKH, 1962, 1966). There is a recursive quasi-well-ordering A such that $\mathsf{T} \exp \mathsf{A}$ is not a quasi-well-ordering where $\mathsf{T} \in 2$.

PROOF. (i) is well-known.

(ii) Assume A is a quasi-well-ordering, B is a well-ordering and there is a recursive descending chain in $\mathsf{A} \exp \mathsf{B}$, namely

$$\{c_i\}_{i=0}^{\infty}, \quad \text{where} \quad c_i = e\begin{pmatrix} b_{i0} \dots b_{in_i} \\ a_{i0} \dots a_{in_i} \end{pmatrix}.$$

Now let b_{ir_i} be defined for each i by

$$r_i = \mu_s\{b_{i+1,s} <_{\mathsf{B}} b_{i,s}\} \quad \text{or } n_i \quad \text{if there is no such } s.$$

Let $e_i = b_{ir_i}$ and let $\{f_i\}$ be the sequence of distinct e_i. Now $\{f_i\}$ is finite, $= \{f_0, \dots, f_n\}$, say, since otherwise $\{f_i\}$ is a descending chain in B which is impossible. It follows that if $f_n = e_s$ then we have $e_t = f_n$ for $t \geq s$. Now for such t we have

$$c_t = e\begin{pmatrix} b_0 \dots b_{s-1} & e_s \\ a_{t0} \dots a_{t,s-1} & a_{t,s} \end{pmatrix},$$

where s and $b_0, \dots, b_{s-1}, e_s$ are fixed. Since the c_t form an (infinite) recursive descending chain one of the sequences

$$\{a_{r0}\}_{r=t}^{\infty}, \dots, \{a_{rs}\}_{r=t}^{\infty}$$

must be infinite, say $\{a_{rq}\}_{r=t}^{\infty}$. This sequence is clearly recursive since it only depends on two fixed numbers t, q and it is the sequence

$$\{k((c_r)_q)\}_{r=t}^{\infty}.$$

Hence A contains a recursive descending chain which is impossible.

(iii) By KLEENE (1952) theorem IV we can obtain all the sets E_n which can be expressed in Σ_2 form

$$\{a : (\exists x)(\forall y)\, \mathrm{R}(a, x, y)\}$$

in a uniform way by setting

$$\mathrm{E}_n = \{a : (\exists x)(\forall y)\, \mathrm{T}_2(n, a, x, y)\},$$

where T_2 is a fixed (primitive) recursive predicate. Now we can define a set S which is a Σ_2 set but whose complement, though infinite, contains no infinite Σ_2 subset, i.e. S is a "Σ_2-simple" set. We shall show that any recursive descending chain in an ordering B we define below would compel an infinite Σ_2 set to lie in $\bar{S}$ which is impossible. Let

$$S = \{a:(\exists n)(\exists x)\{(\forall y)\, T_2(n, a, x, y) \;\&\; 2n < a$$
$$\& (\forall a')(\forall x')[j(a', x') < j(a, x) \;\&\; 2n < a' \Rightarrow (\exists y) \neg T_2(n, a', x', y)]\}\},$$

i.e. $a \in S$ if, and only if, a is the "first" element of some E_n (as determined by the ordering from the $j(a, x)$) and $a > 2n$.

Clearly, if $a \in S$, then $a \in E_n$ for some n such that $a > 2n$. Secondly, if E_n is infinite, then $E_n \cap S \neq \emptyset$. For if E_n is infinite, there is some element $b \in E_n$ with $b > 2n$ and we choose the smallest, i.e. take

$$a = k(\mu_c\{k(c) \in E_n \;\&\; k(c) > 2_n \;\&\; (\forall y)\, T_2(n, k(c), l(c), y)\}).$$

Then a is well-defined and by the definition of S, $a \in S$. Hence $\bar{S}$ contains no infinite Σ_2 subset.

Next $\bar{S}$ is infinite for of the integers $1, \ldots, 2n+1$, at most n are in S so at least one of $n+1, \ldots, 2n+1$ is in $\bar{S}$.

Now let $C = \bar{S}$ so

$$C = \{a:(\forall x)(\exists y)\, Q(a, x, y)\}$$

for some (primitive) recursive Q. Let us call *a k-attained by n* if

$$(\forall x)_{<k}(\exists y)_{<n}\, Q(a, x, y)$$

and call *a k-attained* if, for some n, a is k-attained by n. So $x \in C \Leftrightarrow x$ is k-attained for every k. Now we define a subset B of $j(\mathcal{N}, 2^0 * \text{Seq})$ by

$$x \in B \Leftrightarrow x = j(m, \langle x_0, \ldots, x_k \rangle)$$
$$\& \; m = \mu_r\{(\forall i)(0 < i \leq r \Rightarrow x_i \text{ is } k\text{-attained by } r)\}$$
$$\& \; x_0 = 0 \;\&\; (\forall i)_{<k}(x_i < x_{i+1} \leq 2x_i + 1).$$

We also write

$$x_i = [x]_i \quad \text{for} \quad i = 0, \ldots, k.$$

Now we order B by stipulating

$$\langle x, y \rangle \in \mathsf{B} \Leftrightarrow x, y \in B \;\&\; (l(x) \succcurlyeq l(y) \quad \text{or} \quad l(y) = 2^0).$$

Clearly B is a recursive linear ordering. Suppose $\{a_n\}$ is a recursive descending chain in B. We shall prove that this implies C contains an infinite Σ_2 subset. By definition of B, $[a_n]_0 = 0$ for all n. Suppose that for $n > N$,

$$[a_n]_0, \ldots, [a_n]_r$$

are constant. Now

$$a_n >_{\mathsf{B}} a_{n+1},$$

hence

$$[a_n]_{r+1} \leq [a_{n+1}]_{r+1} \quad \text{by the definition of } \preccurlyeq$$

and for $n > N$

$$[a_n]_r < [a_n]_{r+1} \leq 2[a_n]_r + 1,$$

so

$$\{[a_n]_{r+1}\}$$

is a bounded non-decreasing sequence of integers and so is constant for $n > N'$, where N' is some integer $> N$. Therefore

$$[a_n]_0, \ldots, [a_n]_{r+1}$$

are constant for $n > N'$. Hence for each r, $[a_n]_r$ is eventually constant. We conclude that the sequences represented by sequence numbers $l(a_n)$ must be unbounded in length, i.e. each $[a_n]_r$ is k-attained for arbitrarily large k and hence for all k, provided that n is sufficiently large. Hence

$$\mathrm{E}' = \{a' : (\exists i)\,(\exists n)\,(\forall m)\,[m > n \Rightarrow [a_m]_i = a']\}$$

is an infinite Σ_2 set contained in C. This contradicts the definition of C and we conclude that B is a (recursive) quasi-well-ordering.

By theorem 6.1.2 it follows that $\mathsf{A} = \mathsf{W} \cdot \mathsf{B}$ is also a (recursive) quasi-well-ordering. Now let

$$\mathsf{T} = \{\langle 0, 0\rangle, \langle 0, 1\rangle, \langle 1, 1\rangle\},$$

then $\mathsf{T} \in 2$ and if $x \in \mathrm{C}‘\mathsf{T} \exp \mathsf{A}$ then x is of the form

$$x = e \begin{pmatrix} a_0 \ldots a_k \\ 1 \ldots 1 \end{pmatrix},$$

where

$$a_0 >_{\mathsf{A}} a_1 \cdots >_{\mathsf{A}} a_k .$$

We abbreviate the notation to

$$x = (a_0, \ldots, a_k) .$$

Now let g be the one-one recursive function which enumerates the (infinite recursive) set B in order of magnitude. Suppose

$$j(m, \langle x_0, \ldots, x_{k-1}\rangle), j(q, \langle x_0, \ldots, x_k\rangle) \in \mathrm{B},$$

then

$$j(m, \langle x_0, \ldots, x_{k-1}\rangle) <_{\mathrm{B}} j(q, \langle x_0, \ldots, x_k\rangle), \tag{2}$$

since if $x_0, \ldots, x_k$ are k-attained by q then $x_0, \ldots, x_{k-1}$ are $(k-1)$-attained by some $m \leq q$. So in particular the smallest number in B must be of the form

$$g(0) = j(\mathrm{N}, \langle 0\rangle) \quad \text{for some} \quad \mathrm{N}.$$

We now define a function f which we shall show yields a recursive descending chain in $\mathsf{T}\exp\mathsf{A}$.

Set

$$f(0) = j(2, j(\mathrm{N}, \langle 0\rangle))$$

and suppose for $n \geq 0$,

$$f(n) = (a_0, \ldots, a_k),$$
$$g(n) = j(m, \langle x_0, \ldots, x_1\rangle).$$

We describe a procedure for obtaining $f(n+1)$ and only afterwards show that $f(n+1)$ always exists. It will be clear that f is partial recursive. Let a_i be the first term a_j in $f(n)$ of the form

$$j(p, j(q, \langle x_0, \ldots, x_{l-1}\rangle)) \quad \text{with} \quad p \geq 1. \tag{3}$$

Let

$$a_i' = j(p-1, j(q, \langle x_0, \ldots, x_{l-1}\rangle))$$

and replace a_i by a_i' to obtain

$$(a_0, \ldots, a_{i-1}, a_i', a_{i+1}, \ldots, a_k), \tag{4}$$

unless $a_i' = a_{i+1}$ when we omit a_{i+1}.

Next take the term

$$a' = j(x_1 + 2, j(m, \langle x_0, \ldots, x_1\rangle)) \tag{5}$$

and insert it in the sequence (4) to obtain

$$f(n+1) = (a_0, \ldots, a_{i-1}, a_i', a_{i+1}, \ldots, a_r, a', a_{r+1}, \ldots, a_k),$$

where

$$a_0 >_{\mathsf{A}} \cdots >_{\mathsf{A}} a_{i-1} >_{\mathsf{A}} a_i' \; (>_{\mathsf{A}} a_{i+1}) >_{\mathsf{A}} \cdots >_{\mathsf{A}} a_k.$$

(That this is the correct order follows from (6) below.) Now

$$j(m, \langle x_0, \ldots, x_l\rangle)) <_{\mathrm{B}} j(q, \langle x_0, \ldots, x_{l-1}\rangle),$$

so

$$j(x_l+2, j(m, \langle x_0, \ldots, x_l\rangle)) <_{\mathsf{A}} j(p-1, j(q, \langle x_0, \ldots, x_{l-1}\rangle)) \qquad (6)$$

and

$$f(n+1) <_{\mathsf{A}\exp\mathsf{T}} f(n).$$

It remains to show that $f(n)$ is always defined. Clearly, $f(0)$ is well-defined and contains a term of the form (3), namely $j(2, j(N, \langle 0\rangle))$, since $lg(1)$ is a proper extension of the sequence number $lg(0)$. Now suppose that $f(n)$ contains such a term, then by (2)

$$g(n) = j(m, \langle x_0, \ldots, x_l\rangle),$$

where for some a_i in $f(n)$

$$a_i = j(p, j(q, \langle x_0, \ldots, x_{l-1}))\rangle.$$

Now the p in this a_i was obtained in one of two ways: either (i) as x_l+2 in the construction of the a' in the previous step or (ii) as $p-1$ in the construction of a_i'. In case (i), clearly $p>1$ and in case (ii) we observe that x_l satisfies

$$x_{l-1} < x_l \leq 2x_{l-1} + 1,$$

so there are at most $x_{l-1}+1$ choices for x_l. Thus case (ii) could only have arisen at most $x_{l-1}+1$ times in the calculations of values of f which used this particular p. In either case, therefore, $p \geq 1$. It follows that $f(n+1)$ is well-defined and contains a term a' of the form (2) by (6). Hence by induction on n, $f(n)$ is defined for all n so f is partial recursive and total, hence recursive. This completes the proof.

7.3.

7.3.1 Definition. If $A \geq 1$ and $B \neq 0$ then $A^B = A \exp B = \mathrm{COT}(\mathsf{A} \exp \mathsf{B})$ for $\mathsf{A} \in A$ and $\mathsf{B} \in B$. $0^A = 0$ if $A \neq 0$; $0^0 = 1$, $A^0 = 1$ if $A \geq 1$.

By theorem 7.1.6, $A \exp B$ is uniquely defined.

7.3.2 Theorem. (i) If $A \neq 0$ and B are co-ordinals, then $A \exp B$ is always defined and is a co-ordinal,

(ii) There is a recursive quord A such that 2^A is defined and is not a quord.

(iii) If $A \exp B$ is a quord (co-ordinal) and $A \neq 0, 1$ and $B \neq 0$ then A and B are quords (co-ordinals).

PROOF. (i), (ii) are immediate from theorem 7.2.1.(i), (iii).

(iii) Since $A \neq 0, 1$, if $\mathsf{A} \in A$, then there exists $a \in C`\mathsf{A}$ such that $a \neq \min(\mathsf{A})$. Hence the map

$$b(x) = e\begin{pmatrix} x \\ a \end{pmatrix}$$

is a one-one partial recursive map of $C`\mathsf{B}$ into $C`\mathsf{A} \exp \mathsf{B}$ for $\mathsf{B} \in B$. Hence if $\{d_n\}$ is a (recursive) descending chain in B, then $\{b(d_n)\}$ is a (recursive) descending chain in $\mathsf{A} \exp \mathsf{B}$. Hence if $A \exp B$ is a quord (co-ordinal), then B is a quord (co-ordinal).

Similarly, if $B \neq 0$ and $\mathsf{B} \in B$ then there exists $b \in C`\mathsf{B}$ and the map

$$a(x) = e\begin{pmatrix} b \\ x \end{pmatrix} \quad \text{if} \quad x \neq \min(\mathsf{A}),$$
$$\text{is undefined otherwise,}$$

is a one-one partial recursive map of $C`\mathsf{A}(\min \mathsf{A}$ into $C`\mathsf{A} \exp \mathsf{B}$ for $\mathsf{A} \in A$. Hence if $\{d_n\}$ is a (recursive) descending chain in A, then $\{a(d_n)\}$ is a (recursive) descending chain in $\mathsf{A} \exp \mathsf{B}$. Hence if $A \exp B$ is a quord (co-ordinal) then A is a quord (co-ordinal).

We recall that A^B is defined only if $A=0$ or $A \geq 1$.

7.3.3 THEOREM. $(A^B)^C = A^{BC}$.

PROOF. If $A=0$, $B=C=1$ then both sides of the equation are 0; in all other cases for A, B, $C=0$ or 1, both sides are 1. We therefore assume $A, B, C \neq 0, 1$. Let $\mathsf{A} \in A$, $\mathsf{B} \in B$ and $\mathsf{C} \in C$ where $\mathsf{A}, \mathsf{B}, \mathsf{C} \subseteq \mathsf{R}$, then *assuming that a, b, c below* (with or without sub- or superscripts) *always denote elements belonging respectively to* $C`\mathsf{A}$, $C`\mathsf{B}$ *and* $C`\mathsf{C}$,

$$(\mathsf{A}^\mathsf{B})^\mathsf{C} = \{\langle e(D), e(D')\rangle : D = \begin{pmatrix} c_0 \dots c_n \\ q_0 \dots q_n \end{pmatrix} \in (\mathsf{A}^\mathsf{B}, \mathsf{C})$$

$$\& \; D' = \begin{pmatrix} c'_0 \dots c'_{n'} \\ q'_0 \dots q'_{n'} \end{pmatrix} \in (\mathsf{A}^\mathsf{B}, \mathsf{C}) \; \& \; (\forall r)\, (q_r = e(Q_r) \; \& \; q'_r = e(Q'_r)$$

$$\& \; Q_r = \begin{pmatrix} b_{r0} \dots b_{rt_r} \\ a_{r0} \dots a_{rt_r} \end{pmatrix} \in (\mathsf{A}, \mathsf{B}) \; \& \; Q'_r = \begin{pmatrix} b'_{r0} \dots b'_{rt'_r} \\ a'_{r0} \dots a'_{rt'_r} \end{pmatrix} \in (\mathsf{A}, \mathsf{B})$$

$$\begin{aligned}&\&\colon D=\mathbf{0}\,.\vee.\,D\neq\mathbf{0}\ \&\ [(n\le n'\ \&\ (\forall r)(r\le n\Rightarrow c_r=c'_r\ \&\ q_r=q'_r))\\&\quad\vee(\exists r)(\forall t)\{(t<r\Leftrightarrow(\forall s\le t)(c_s=c'_s\ \&\ q_s=q'_s))\\&\qquad\&\ (c_r\prec c'_r\,.\vee.\,c_r=c'_r\ \&\ \langle q_r,q'_r\rangle\in\mathsf{A}^{\mathsf{B}})\}]\}.\end{aligned}$$

Now

$$\begin{aligned}\langle q_r,q'_r\rangle\in\mathsf{A}^{\mathsf{B}}\Leftrightarrow&(t_r\le t'_r\ \&\ (\forall s)(s\le t_r\Rightarrow b_{rs}=b'_{rs}\ \&\ a_{rs}=a'_{rs}))\\&\vee(\exists u)(\forall v)\{(v<u\Rightarrow b_{rv}=b'_{rv}\ \&\ a_{rv}=a'_{rv})\\&\qquad\&\ [b_{ru}\prec b'_{ru}\,.\vee.\,b_{ru}=b'_{ru}\ \&\ a_{ru}<a'_{ru}]\}.\end{aligned}$$

On the other hand

$$\begin{aligned}\mathsf{A}^{\mathsf{BC}}=\{\langle e(D),e(D')\rangle\colon D=&\begin{pmatrix}j(b_0,c_0)\dots j(b_n,c_n)\\a_0\quad\dots\quad a_n\end{pmatrix}\in(\mathsf{A},\mathsf{BC})\\&\&\ D'=\begin{pmatrix}j(b'_0,c'_0)\dots j(b'_{n'},c'_{n'})\\a'_0\quad\dots\quad a'_{n'}\end{pmatrix}\in(\mathsf{A},\mathsf{BC})\\&\&\colon D=\mathbf{0}\,.\vee.\,D\neq\mathbf{0}\ \&\ [(n\le n'\ \&\ (\forall r)(r\le n\Rightarrow j(b_r,c_r)\\&=j(b'_r,c'_r)\ \&\ a_r=a'_r))\vee(\exists r)(\forall t)\{(t<r\Leftrightarrow(\forall s\le t)(j(b_s,c_s)\\&=j(b'_s,c'_s)\ \&\ a_s=a'_s))\ \&\ (\langle j(b_r,c_r),j(b'_r,c'_r)\rangle\in\mathsf{B}\cdot\mathsf{C}\\&\&\ j(b_r,c_r)\neq j(b'_r,c'_r)\,.\vee.\,j(b_r,c_r)=j(b'_r,c'_r)\ \&\ a_r\preccurlyeq a'_r)\}]\}.\end{aligned}$$

But

$$\langle j(b_r,c_r),j(b'_r,c'_r)\rangle\in\mathsf{B}\cdot\mathsf{C}\Leftrightarrow c_r\prec c'_r\vee(c_r=c'_r\ \&\ b_r\preccurlyeq b'_r)$$

and

$$j(b_r,c_r)=j(b'_r,c'_r)\Leftrightarrow b_r=b'_r\ \&\ c_r=c'_r.$$

Now let p be the partial recursive function defined only on the recursive set

$$\{e(G)\colon G=\begin{pmatrix}i_0\dots i_k\\e_0\dots e_k\end{pmatrix}\ \&\ (\forall i)(e_i\in\rho e)\},$$

by $p(0)=0$,

$$\begin{aligned}&p\left(e\begin{pmatrix}c_0&\dots&c_n\\e(Q_0)&\dots&e(Q_n)\end{pmatrix}\right)=\\&\quad=e\begin{pmatrix}j(b_{00},c_0)\dots j(b_{0m_0},c_0)\,j(b_{1,0},c_1)\dots j(b_{1,m_1},c_1)\dots j(b_{n0},c_n)\dots j(b_{nm_n},c_n)\\a_{00}\quad\dots\quad a_{0m_0}\qquad a_{1,0}\quad\dots\quad a_{1,m_1}\quad\dots\quad a_{n0}\quad\dots\quad a_{nm_n}\end{pmatrix},\end{aligned}$$

where

$$Q_r=\begin{pmatrix}b_{r0}\dots b_{rm_r}\\a_{r0}\dots a_{rm_r}\end{pmatrix}.$$

Using the definitions of $(\mathsf{A}^{\mathsf{B}})^{\mathsf{C}}$ and A^{BC} given above the reader will readily verify that p is order preserving. That p is one-one and partial recursive follows easily from theorem 7.1.2. It follows that

$$p:(\mathsf{A}^{\mathsf{B}})^{\mathsf{C}} \simeq \mathsf{A}^{\mathsf{BC}}.$$

Taking C.O.T.s completes the proof.

7.3.4 THEOREM. $A \exp(B+C)=A^B \cdot A^C$.

PROOF. Let $\mathsf{A}\in A$, $\mathsf{B}\in B$ and $\mathsf{C}\in C$ where $\mathsf{B})\,(\mathsf{C}$. Then $\mathsf{B} \dotplus \mathsf{C}$ is well-defined and by definition 2.1.1 there exist disjoint r.e. sets B, C such that

$$C`\mathsf{B} \subseteq \mathrm{B} \quad \text{and} \quad C`\mathsf{C} \subseteq \mathrm{C}.$$

Now

$$C`(\mathsf{A} \exp(\mathsf{B} \dotplus \mathsf{C})) = \{e(\boldsymbol{D}): \boldsymbol{D} \in (\mathsf{A}, \mathsf{B} \dotplus \mathsf{C})\}$$

and

$$\boldsymbol{D} \in (\mathsf{A}, \mathsf{B} \dotplus \mathsf{C}) \Leftrightarrow \boldsymbol{D} = \begin{pmatrix} e_0 \cdots e_r \\ a_0 \cdots a_r \end{pmatrix} \;\&\; (\forall i)_{\leq r}(a_i \mp \min(\mathsf{A}))$$
$$\& \; e_0 >_{\mathsf{B}+\mathsf{C}} e_1 \cdots >_{\mathsf{B}+\mathsf{C}} e_r .$$

This last clause is equivalent to

$$\begin{aligned} & e_0 >_{\mathsf{B}} \cdots e_r \\ \vee\; & e_0 >_{\mathsf{C}} \cdots e_r \\ \vee\; & (\exists s)\,(s < r \;\&\; e_0 >_{\mathsf{C}} \cdots e_s \;\&\; e_{s+1} >_{\mathsf{B}} \cdots e_r). \end{aligned}$$

On the other hand

$$C`(\mathsf{A}^{\mathsf{B}} \cdot \mathsf{A}^{\mathsf{C}}) = \{j(e(\boldsymbol{D}_1), e(\boldsymbol{D}_2)): \boldsymbol{D}_1 \in (\mathsf{A}, \mathsf{B}) \;\&\; \boldsymbol{D}_2 \in (\mathsf{A}, \mathsf{C})\}$$

and

$$\boldsymbol{D}_1 \in (\mathsf{A}, \mathsf{B}) \Leftrightarrow \boldsymbol{D}_1 = \begin{pmatrix} e_{s+1} \cdots e_r \\ a_{s+1} \cdots a_r \end{pmatrix} \;\&\; (\forall i)_{<r-s}(a_{s+i+1} \mp \min(A))$$
$$\& \; e_{s+1} >_{\mathsf{B}} \cdots e_r$$

and

$$\boldsymbol{D}_2 \in (\mathsf{A}, \mathsf{C}) \Leftrightarrow \boldsymbol{D}_2 = \begin{pmatrix} e_0 \cdots e_s \\ a_0 \cdots a_s \end{pmatrix} \;\&\; (\forall i)_{\leq s}\,(a_i \mp \min(\mathsf{A})) \;\&\; e_0 >_{\mathsf{C}} \cdots e_s.$$

Recalling that if $r=s$, then $\boldsymbol{D}_1=\boldsymbol{0}$ and $e(\boldsymbol{D}_1)=0$, and that if $s=-1$, $\boldsymbol{D}_2=\boldsymbol{0}$ and $e(\boldsymbol{D}_2)=0$, it follows that the map p defined by

$$p(x) = l(x) * k(x)$$

is order-preserving between $\mathsf{A}^{\mathsf{B}} \cdot \mathsf{A}^{\mathsf{C}}$ and $\mathsf{A}\exp(\mathsf{B} \dotplus \mathsf{C})$. Now let

$$\mathrm{D} = \{j(e(\boldsymbol{D}_1), e(\boldsymbol{D}_2)) : \boldsymbol{D}_1 = \begin{pmatrix} b_0 \dots b_m \\ a_0 \dots a_m \end{pmatrix} \ \&\ \boldsymbol{D}_2 = \begin{pmatrix} c_0 \dots c_n \\ a'_0 \dots a'_n \end{pmatrix}$$
$$\&\ (\forall i)\,(a_i, a'_i \neq \min(\mathsf{A})\ \&\ b_i \in \mathrm{B}\ \&\ c_i \in \mathrm{C})\}$$

and let q be the map p with domain restricted to D. Then q is clearly partial recursive since D is r.e. Further q is one-one, for suppose $q(x) = q(y) = z$, say. Then

$$z = pj(e(\boldsymbol{D}_1), e(\boldsymbol{D}_2)) = e(\boldsymbol{D})$$

for some bracket symbols $\boldsymbol{D}_1$, $\boldsymbol{D}_2$, where $\boldsymbol{D}$ is uniquely determined since $\boldsymbol{e}$ is one-one. Now

$$\boldsymbol{D} = \begin{pmatrix} e_0 \dots e_r \\ a_0 \dots a_r \end{pmatrix},$$

where $a_i \neq \min(\mathsf{A})$ and $e_i \in \mathrm{B} \cup \mathrm{C}$. Further, there is precisely one number s such that

$$-1 \leq i \leq s \Rightarrow e_i \in \mathrm{C}\ \&\ s < i \leq r \Rightarrow e_i \in \mathrm{B}$$

by the definition of the domain of q. Therefore $\boldsymbol{D}_1$, $\boldsymbol{D}_2$ are uniquely determined and q is one-one. Thus q is a recursive isomorphism from $\mathsf{A}^{\mathsf{B}} \cdot \mathsf{A}^{\mathsf{C}}$ onto $\mathsf{A}\exp(\mathsf{B} \dotplus \mathsf{C})$; taking C.O.T.s completes the proof.

7.3.5 COROLLARY.

$$A^n = \underbrace{A \cdot \dots \cdot A}_{n \text{ factors}}.$$

PROOF. By induction on n is left to the reader.

7.3.6 THEOREM. (i) $A^n \cdot A^W = A^W$,

(ii) $(A^n)^W = A^W$.

PROOF. By theorem 6.2.2. (iv), (vi).

7.4 We now prove an analogue of lemma 6.3.1.[72] Since, if $A \neq 0$ is a recursive quord, then $1 \leq A$ (corollary 3.3.2) the following lemma holds for recursive quords and co-ordinals.

7.4.1 LEMMA. If $A \geq 1$ is a quord, then

$$BA = A \Leftrightarrow (\exists C)\,(B^W C = A).$$

PROOF. $B^W C = A \Rightarrow BA = B^{1+W} C = B^W C = A$ by theorems 7.3.6 and 5.2.6.(i).

Conversely, suppose $BA = A$. We may assume $A > 1$ since otherwise we may take $C = 1$ $(= B)$. By hypothesis there exist quasi-well-orderings $\mathsf{A}, \mathsf{B} \subseteq \mathsf{R}$ and a recursive isomorphism f such that

$$\mathsf{A} \in A, \mathsf{B} \in B \quad \text{and} \quad f : \mathsf{A} \simeq \mathsf{BA}.$$

Let $\mathrm{A} = C`\mathsf{A}$, $\mathrm{B} = C`\mathsf{B}$. Since $1 < A$ we may assume

$$\min(\mathsf{A}) = \min(\mathsf{B}) = 0.$$

Now we put (even if $a \notin \mathrm{A}$, though then a_r may be undefined for all r greater than some s)

$$a_0 = a, \quad a_{r+1} = lf(a_r).$$

Then, if $a \in \mathrm{A}$,

$$f(a) = j(b_0, a_1) \quad \text{for some} \quad b_0 \in \mathrm{B},\ a_1 \in \mathrm{A} \tag{7}$$

and in general

$$f(a_r) = j(b_r, a_{r+1}) \quad \text{for some} \quad b_r \in \mathrm{B},\ a_{r+1} \in \mathrm{A}.$$

We first of all show that $a_i \succcurlyeq a_{i+1}$ for all i if $a \in \mathrm{A}$. Suppose $a \prec a_1$, then

$$j(0, a) <_{\mathsf{BA}} j(0, a_1)$$

and hence

$$f^{-1} j(0, a) = a' \prec f^{-1} j(0, a_1) \preccurlyeq a \quad \text{by (7)}.$$

Therefore

$$j(0, a') <_{\mathsf{BA}} j(0, a)$$

and

$$a'' = f^{-1} j(0, a') \prec f^{-1} j(0, a) = a' \quad \text{and so on}.$$

Since, for $a \in \mathrm{A}$, $a', a'', \ldots$ are all defined, $\{a, a', a'', \ldots\}$ is a recursive descending chain in A which is impossible since A is a quord.

Now suppose $a_i = a_{i+1}$ for some $a \in \mathrm{A}$ and some i, then

$$f(a_i) = j(b_i, a_{i+1}) \quad \text{and} \quad j(0, a_{i+1}) \leq_{\mathsf{BA}} j(b_i, a_{i+1}).$$

Now if $b_i \neq 0$ then (say)

$$a' = f^{-1} j(0, a_{i+1}) <_{\mathsf{A}} f^{-1} j(b_i, a_{i+1}) = a_{i+1}.$$

Hence, as above, we get a recursive descending chain since f is order preserving. This is impossible so we conclude $b_i = 0$. Thus we have $a = a_0 \succ a_1 \succ \cdots \succ a_r = a_{r+1} = \cdots$ and $b_r = 0 = b_{r+1} = \cdots$.

Let n be the partial recursive function defined by

$$n(a) = \mu_r \{a_r = a_{r+1}\},$$

then n is always defined if $a \in \mathrm{A}$ and moreover, $\lambda ai\, a_i$ and $\lambda bi\, b_i$ are partial recursive functions since the above process of calculating the a_i and b_i is clearly effective.

Let $\mathsf{C} = \mathsf{A}[\{x : lf(x) = x\}$ and let $\mathsf{D} = g(\mathsf{B}^{\mathsf{W}})$ where g is the partial recursive function defined only on ρe, which maps only bracket symbol images of the form

$$e\begin{pmatrix} n_0 n_1 \dots n_s \\ b_{n_0} b_{n_1} \dots b_{n_s} \end{pmatrix} \quad \text{where} \quad n_i, b_{n_i} \in \mathcal{N} \quad \text{and} \quad n_0 > n_1 > \cdots > n_s \geq 0$$

onto

$$e\begin{pmatrix} n_0 & n_0 - 1 \dots n_1 + 1 & n_1 & n_1 - 1 \dots n_s & n_s - 1 \dots 0 \\ b_{n_0} & 0 \quad \dots \quad 0 & b_{n_1} & 0 \quad \dots b_{n_s} & 0 \quad \dots 0 \end{pmatrix}.$$

(Note. Although this last expression is not technically a bracket symbol since we are allowing zeros in the bottom row, nevertheless no confusion will arise since *all* natural numbers $\leq n_0$ occur if the bracket symbol is not **0** and if it is g gives the value 0. Further, $n_0 > 0$ or $b_{n_0} > 0$ since $n_0 = b_{n_0} = 0$ only if we are dealing with the minimum element and this maps to 0.)

I.e. $g(x)$ inserts the missing positive integers in the top row of $e^{-1}(x)$ and in the columns where an integer was missing inserts $\min(\mathsf{B})$ in the bottom row and finally g takes the image under e of the resulting bracket symbol $(g(0) = 0)$.

It is clear that g, and hence D, is well-defined.

We shall now show that $\mathsf{A} \simeq \mathsf{D} \cdot \mathsf{C}$ from which it follows at once that $A = B^W C$ where $C = \mathrm{COT}(\mathsf{C})$. Let

$$h(x) = j\left(e\begin{pmatrix} n(x) - 1 & \dots & i & \dots & 0 \\ kf(lf)^{n(x)-1}(x) & \dots & kf(lf)^i(x) & \dots & kf(x) \end{pmatrix}, (lf)^{n(x)}(x)\right).$$

Clearly, h is partial recursive. Suppose $h(x) = h(y)$, then $kh(x) = kh(y)$ and $lh(x) = lh(y)$. Hence

$$(lf)^{n(x)}(x) = (lf)^{n(y)}(y). \tag{8}$$

Now, since e is one-one we have

$$n(x) = n(y)$$

and therefore, for $0 \leq r < n(x)$,

$$kf(lf)^r(x) = kf(lf)^r(y). \tag{9}$$

Putting $r = n(x) - 1$ in (9) and using (8) we have

$$f(lf)^{n(x)-1}(x) = f(lf)^{n(x)-1}(y).$$

But f is one-one, hence

$$(lf)^{n(x)-1}(x) = (lf)^{n(x)-1}(y).$$

Now assume

$$(lf)^s(x) = (lf)^s(y), \tag{10}$$

where $s < n(x) = n(y)$. Then by (9) with $r = s - 1$,

$$(kf)(lf)^{s-1}(x) = (kf)(lf)^{s-1}(y)$$

and using (10)

$$f(lf)^{s-1}(x) = f(lf)^{s-1}(y).$$

By the one-one property of f

$$(lf)^{s-1}(x) = (lf)^{s-1}(y)$$

and by induction it follows that $x = y$, i.e. h is one-one.

We next prove that h is order-preserving. Suppose $a <_{\mathsf{A}} a'$ and let $n =$ maximum of $n(a)$ and $n(a')$ (which are both defined since $a, a' \in \mathsf{A}$). Since f is order-preserving between A and BA,

$$\begin{aligned} a_i <_{\mathsf{A}} a_i' \Leftrightarrow f(a_i) <_{\mathsf{BA}} f(a_i') & \\ \Leftrightarrow a_{i+1} <_{\mathsf{A}} a_{i+1}' \quad & \text{or} \\ a_{i+1} = a_{i+1}' \;\&\; b_{i+1} <_{\mathsf{B}} b_{i+1}', & \end{aligned}$$

where $b_{i+1} = kf(a_{i+1})$. Hence

$$\begin{aligned} a <_{\mathsf{A}} a' \Leftrightarrow\; & a_n <_{\mathsf{A}} a_n' \quad \text{or} \\ & a_n = a_n' \;\&\; b_n <_{\mathsf{A}} b_n' \quad \text{or} \\ & a_n = a_n' \;\&\; a_{n-1} = a_{n-1}' \;\&\; b_{n-1} <_{\mathsf{B}} b_{n-1}' \quad \text{or} \\ & \cdots\cdots \\ & \cdots\cdots \\ & \cdots\cdots \quad \text{or} \\ & a_n = a_n' \;\&\cdots\&\; a_1 = a_1' \;\&\; b_1 <_{\mathsf{B}} b_1' \end{aligned} \tag{11}$$

$$\begin{aligned}\Leftrightarrow a_n <_{\mathsf{A}} a'_n \quad &\text{or}\\ a_n = a'_n \;\&\; b_n <_{\mathsf{B}} b'_n \quad &\text{or}\\ \dots\dots\\ \dots\dots\\ \dots\dots \quad &\text{or}\\ a_n = a'_n \;\&\; b_n = b'_n \;\&\cdots\&\; b_2 = b'_2 \;\&\; b_1 <_{\mathsf{B}} b'_1\,,\end{aligned}$$

since $f(a_i) = j(a_{i+1}, b_{i+1})$.

Now

$$h(a) <_{\mathsf{DC}} h(a') \Leftrightarrow a_n <_{\mathsf{A}} a'_n \quad \text{or}$$
$$a_n = a'_n \;\&\; (\exists r)\,(1 \le s < r \Rightarrow b_s = b'_s \;\&\; b_r <_{\mathsf{B}} b'_r)\,. \qquad (12)$$

(11) and (12) are equivalent since, as we observed above, if $n > n(a)$, then $a_n = a_{n(a)}$ and $b_n = \min(\mathsf{B}) = 0$.

Finally, h maps C‘**A** onto C‘**DC** for suppose

$$x = j\left(e\begin{pmatrix} n \dots 0 \\ b_n \dots b_0 \end{pmatrix}, c\right) \in \text{C‘}\mathsf{DC}\,,$$

where $c \in$ C‘**C** (and $b_r \in$ B). Let $a_0 = c$, $a_{r+1} = f^{-1}(j(b_r, a_r))$. Then $a_0 \in$ A and if $a_r \in$ A, then $a_{r+1} \in$ A; in particular $a_n \in$ A and $a_n = h^{-1}(x)$. We have therefore proved

$$h : \mathsf{A} \simeq \mathsf{DC}$$

and the lemma is established.

7.4.2 THEOREM. $n^W = W$ if $n > 0$.

PROOF. It suffices to prove that if $\mathsf{N} \in n$, then $\mathsf{W} \simeq \mathsf{N}^{\mathsf{W}}$. Let

$$\mathsf{N} = \{\langle x, y\rangle : 0 \le x \le y < n\}, \quad \text{then clearly} \quad \mathsf{N} \in n\,.$$

If $s \in \mathcal{N}$, then s is expressible in the form

$$s = n^r a_r + n^{r-1} a_{r-1} + \cdots + a_0\,,$$

where for all i, $0 \le a_i < n$. Let f be the (partial) recursive function defined by

$$f(s) = e\begin{pmatrix} r & r-1 \dots 0 \\ a_r & a_{r-1} \;\dots a_0 \end{pmatrix},$$

where columns with lower entry 0 have been omitted.

$$\text{E.g.} \qquad f(n^2 \cdot 3 + n \cdot 0 + 2) = e\begin{pmatrix} 2 & 0 \\ 3 & 2 \end{pmatrix}$$

for $n>3$. Then, if $u, v \in \mathcal{N}$ and $u = n^r a_r + \cdots + a_0$ and $v = n^r b_r + \cdots + b_0$ (where some of the leading coefficients may be zero),

$$u < v \Leftrightarrow (\exists t \leq r)\left((w > t \Rightarrow a_w = b_w) \; \& \; a_t < b_t\right) \tag{13}$$

and

$$u = v \Leftrightarrow (\forall t)\,(a_t = b_t).$$

(We remark that the fact that $a_r, a_{r-1}, \ldots$ and $b_r, b_{r-1}, \ldots$ may be zero does not affect the ordering.)

But the ordering $\leq$ given in (13) is precisely the ordering in N^{W} of the bracket symbol images

$$e\begin{pmatrix} r \ldots 0 \\ a_r \cdots a_0 \end{pmatrix} \quad \text{and} \quad e\begin{pmatrix} r \ldots 0 \\ b_r \ldots b_0 \end{pmatrix},$$

when columns with bottom row zero have been omitted in the bracket symbols. Clearly, f is one-one and hence $f : \mathsf{W} \simeq \mathsf{N}^{\mathsf{W}}$ and the theorem is established.

CHAPTER 8

ARITHMETIC LAWS AND PRINCIPAL NUMBERS

8.1 We now investigate the arithmetic laws of C.O.T.s. Some laws for addition were established in chapter 4 and we first of all strengthen these.

8.1.1 THEOREM. If A, B, C are quords, then

$$B < C \Rightarrow A + B < A + C .$$

PROOF. Immediate from theorem 4.2.3.(v) and corollary 3.2.9.

However, the other obvious analogue of a classical additive law

$$B \leq C \Rightarrow B + A \leq C + A \tag{*}$$

does not hold in general even for co-ordinals. For let $A = W$, $B = 0$ and $C = V$, then (*) implies W and V are comparable which contradicts corollary 5.2.7. However, (*) does hold if A, B, C are less than what we call a *principal number for addition.* Classically, a principal number for addition (otherwise called a γ-number (BACHMANN, 1955, p. 67) or a prime component (SIERPINSKI, 1958, p. 279)) may be defined as an ordinal number γ such that any of the following (equivalent) conditions holds:

$$\alpha < \gamma \Rightarrow \alpha + \gamma = \gamma , \tag{1^c}$$

$$\alpha + \beta = \gamma \Rightarrow \beta = 0 \quad \text{or} \quad \beta = \gamma , \tag{2^c}$$

$$\alpha, \beta < \gamma \Rightarrow \alpha + \beta < \gamma , \tag{3^c}$$

$$\gamma = 0 \text{ or } \gamma = \omega^{\alpha} \quad \text{for some ordinal } \alpha . \tag{4^c}$$

We leave consideration of a recursive analogue of (4^c) until we have dealt with Cantor Normal Forms, only remarking that we shall show that every co-ordinal of the form W^A is a principal number for addition but that these are not the only principal numbers for addition.

8.1.2 THEOREM. If A is a quord ≥ 1 then the following implications hold between (1), (2), (3) below:

$$(1)\Leftrightarrow(2)\ {}^{\neg}\!\underset{\Rightarrow}{\Leftarrow}(3).$$

$$B < A \Rightarrow B + A = A, \tag{1}$$

$$B + C = A \Rightarrow C = 0 \quad \text{or} \quad C = A, \tag{2}$$

$$B, C < A \Rightarrow B + C < A. \tag{3}$$

PROOF. (1)⇒(2). Suppose (1) holds and $B+C=A$. Then if $C\neq 0$, $B<A$ and by (1), $B+A=A$, hence by corollary 3.2.9, $C=A$. (2)⇒(1). Suppose (2) holds and $B<A$, then there is a $C\neq 0$ such that $B+C=A$ and by (2) we have $C=A$, hence (1) holds. (1)⇒(3). Suppose (1) holds and B, $C<A$, then $B+A=A=C+A$ and hence $(B+C)+A=B+(C+A)=B+A=A$. Since $A\neq 0$, $B+C<A$ by corollary 4.2.5.
(3)¬⇒(1). Let $A=V$, then $1<A$, but $1+V\neq V$ hence (1) does not hold. However (3) clearly holds since $B<V$ if B is finite. This completes the proof.

8.1.3 DEFINITION. A co-ordinal (quord) A is said to be a *principal* (*$\mathscr{Q}$-principal*) *number for addition* if $A>1$ and

$$B < A \Rightarrow B + A = A.$$

We write $\mathscr{H}(+)$ [$\mathscr{Q}(+)$] for the collection of all principal ($\mathscr{Q}$-principal) numbers for addition.

Later on we shall show that $A\in\mathscr{Q}(+)$ implies A is comparable with W^n for every n.

8.1.4 THEOREM. If $A\in\mathscr{H}(+)$, then $|A|$ is a limit number.

PROOF. Clearly no finite co-ordinal is a principal number for addition. Suppose $|A|=\lambda+n$ where λ is a limit number and $n>0$. Then by theorem 5.1.6 $B=A-n$ is well-defined, $|B|=\lambda$ and $B<A$. Hence if $n>0$, $B+A=A$ and $\lambda+n+\lambda=\lambda$ which is impossible.

8.1.5 THEOREM. If P is a principal number for addition and A, B, $C<P$ then

$$A < B \Rightarrow A + C \leq B + C.$$

PROOF. By theorem 8.1.2, A, B, $C<P$ imply $A+C<P$ and $B+C<P$. Therefore by theorem 4.2.8, $A+C$ and $B+C$ are comparable. Now classically,

$$\alpha < \beta \Rightarrow \alpha + \gamma \leq \beta + \gamma,$$

hence by theorem 5.3.4 $A+C \leq B+C$.

8.1.6 COROLLARY. If A, $B<P$ and P is a principal number for addition, then $B \leq A+B$. Conversely, if $A+B<P$, then A, $B<P$.

PROOF. Trivially $A<P$ if $A+B<P$. Since P is a principal number $A+B<P$ implies $A+B+P=P$. But $A<P$ so $A+P=P$ and therefore $A+(B+P)=A+P$. Hence $B+P=P$ and $B \leq P$. Finally we cannot have $B=P$ since then $A+B \nless P$.

8.2 We now give a series of results for quords. All the results in this section were in a sense obtained in the context of ordinal algebras (TARSKI, 1956) by Tarski. Many of the proofs are highly derivative from Tarski's proofs. We do not know which, if any, of these results can be extended to arbitrary C.O.T.s though a start on this problem was made in § 2.4 using Morley's lemma.

8.2.1 THEOREM. If $B \neq 0$ is a quord and there exist C, D such that

$$B + C = A \cdot W = D + B, \quad \text{then} \quad B = A \cdot W.$$

PROOF. By theorem 6.2.5. $D+B=A \cdot W$ implies $A \cdot W \leq^* B$. $B+C=A \cdot W$ implies $B \leq A \cdot W$, hence by theorem 4.2.6, $B=A \cdot W$.

8.2.2 THEOREM. If C is a quord, then

$$A + B + C = C \quad \text{and} \quad (A + B) \cdot W = (B + A) \cdot W \quad \text{iff}$$
$$A + C = B + C = C.$$

PROOF. By lemma 6.3.1, using the first condition,

$$(A + B) \cdot W + D = C \quad \text{for some quord } D.$$

By theorem 6.2.3.(iii),

$$\begin{aligned}(B + A)\cdot W &= B + (A + B)\cdot W \\ &= (A + B)\cdot W \quad \text{by the second condition}.\end{aligned}$$

Hence by lemma 6.3.1,

$$B\cdot W + E = (A + B)\cdot W \quad \text{for some quord } E.$$

Therefore

$$\begin{aligned}B + C &= B + (A + B)\cdot W + D, \\ &= B + (B\cdot W + E) + D, \\ &= B\cdot W + E + D, \qquad \text{by theorem 6.2.2.(iv)} \\ &= (A + B)\cdot W + D = C.\end{aligned}$$

Thus $B+C=C$ and therefore $A+C=A+B+C=C$.

Now suppose $A+C=B+C=C$, then $A+B+C=C$ and $B+A+C=C$. Hence $(B+A)\cdot W+D=C$ and $(A+B)\cdot W+E=C$ for some quords D, E, by lemma 6.3.1. By the directed refinement theorem 2.3.2 it follows that for some F either

$$(B + A)\cdot W + F = (A + B)\cdot W,$$

or

$$(A + B)\cdot W + F = (B + A)\cdot W. \tag{5}$$

We suppose the latter holds. Now $(A + B)\cdot W = A + (B + A)\cdot W$ by theorem 6.2.3.(iii), hence $(B + A)\cdot W \leq^* (A + B)\cdot W$ and hence by (5) and theorem 4.2.6, $(B + A)\cdot W = (A + B)\cdot W$.

This completes the proof.

8.2.3 THEOREM. If A, C (or B, C) are quords, then

$$A + C\cdot n = B + C\cdot n \Rightarrow A + C = B + C.$$

PROOF. By the directed refinement theorem 2.3.2, there exists an E such that either

$$A + E = B \quad \text{or} \quad B + E = A$$

and

$$E + C\cdot n = C\cdot n.$$

Suppose, without loss of generality, that $A+E=B$. By lemma 6.3.1 we have, for some F, $E\cdot W+F=C\cdot n=C+C\cdot(n-1)$ by theorem 6.2.2.(ii). Now by the directed refinement theorem 2.3.2, there exists G such that

$$E\cdot W + G = C \quad \text{and} \quad G + C\cdot(n-1) = F \tag{i}$$

or

$$C + G = E \cdot W \quad \text{and} \quad G + F = C \cdot (n-1). \tag{ii}$$

If (i) holds then

$$\begin{aligned} E + C &= E + E \cdot W + G \\ &= E \cdot W + G \qquad \text{by theorem 6.2.2.(iv)} \\ &= C. \end{aligned}$$

Now $B+C=(A+E)+C=A+(E+C)=A+C$ and the theorem is established for this case.

If (ii) holds, then by theorem 6.2.5 there exists H such that $H+E\cdot W=G$. But then $C\cdot(n-1)=G+F=H+E\cdot W+F=H+C\cdot n=H+C\cdot(n-1)+C$ (last step by theorem 6.2.2.(ii)). By lemma 2.4.8 (proof) we therefore have $C\cdot(n-1)=C\cdot(n-1)+C$. But then by corollary 3.2.9 we have $C=0$ which reduces the theorem to a triviality.

8.2.4 THEOREM. If A, C, D (or B, C) are quords and

$$\begin{aligned} &A + C\cdot n + D = B + C\cdot n, \quad \text{then} \\ &A + C + E = B + C \quad \text{for some (quord) } E. \end{aligned}$$

Before we proceed to the proof we observe that if one sidc (and hence the other) of the first equality is a recursive quord or co-ordinal then the theorem may be stated thus:

$$A + C\cdot n \leq B + C\cdot n \Rightarrow A + C \leq B + C.$$

PROOF (by induction on n). If $n=1$, then the assertion is trivially true with $E=D$. Now suppose the implication holds for n, then

$$A + C\cdot(n+1) + D = B + C\cdot(n+1)$$

implies

$$A + C\cdot n + (C + D) = B + C\cdot n + C.$$

Therefore, by the directed refinement theorem 2.3.2, there is an F such that either

$$A + C\cdot n + F = B + C\cdot n \quad \text{and} \quad C + D = F + C,$$

or

$$A + C\cdot n = B + C\cdot n + F \quad \text{and} \quad F + C + D = C.$$

In the former case, by the induction hypothesis

$$A + C + E = B + C$$

for some E. In the latter case, by theorem 3.2.7, $D=0$ and hence

$$A + C\cdot(n+1) = B + C\cdot(n+1)$$

and by theorem 8.2.3,

$$A + C = B + C.$$

This completes the proof.

8.3 By analogy we now introduce principal numbers for multiplication (cf. BACHMANN, 1955, p. 66).

8.3.1 DEFINITION. A co-ordinal (quord) A is said to be a *principal* *($\mathscr{Q}$-principal) number for multiplication* if $A>2$ and

$$0 < B < A \Rightarrow B \cdot A = A.$$

We write $\mathscr{H}(\cdot)$ [$\mathscr{Q}(\cdot)$] for the collection of all principal ($\mathscr{Q}$-principal) numbers for multiplication.

Later on we shall show that $A \in \mathscr{Q}(\cdot)$ implies A is divisible by or divides W^{W^n} for every n. As in the classical case

$$0 < B < A \Rightarrow BA = A$$

is a stronger condition than

$$BC = A \Rightarrow B = A \quad \text{or} \quad C = A.$$

But also, for co-ordinals, the former condition is stronger than

$$0 < B, C < A \Rightarrow BC < A.$$

For V satisfies this last condition but is not a principal number for multiplication since $2<V$ but by lemma 7.4.1, $2V=V$ implies $2^W=V$ and by theorem 7.4.2, $2^W=W\neq V$.

8.3.2 THEOREM. If A is a co-ordinal in $\mathscr{H}(\cdot)$, then $|A|$ is a limit number.

PROOF. Left to the reader (cf. theorem 8.1.4).

8.3.3 THEOREM. (i) If $B\neq 0$, then $A\leq AB$ when B is a recursive quord, a co-ordinal or $B\geq 1$.

(ii) If $B>1$, then $A<AB$ whenever $A\neq 0$ for an arbitrary quord B.

(iii) If A divides B and $|A|=|B|$, then $A=B$.

PROOF. We prove only (ii) and (iii) leaving (i) to the reader.

(ii) $B>1 \Rightarrow (\mathrm{E}!\, C)\,(B=1+C \,\&\, C \neq 0)$. Hence $AB=A(1+C)=A+AC$, where $AC \neq 0$ if $A \neq 0$; thus $A<AB$.

(iii) By (i) A divides B implies $A \leq B$ hence by theorem 5.3.4, $A=B$.

8.3.4 THEOREM. If A and B are isomorphic well-orderings then there is a unique isomorphism f such that every isomorphism between A and B is an extension of f. (SIERPINSKI, 1958, p. 264, corollary 3).

We now prove a left cancellation law for co-ordinals using this classical theorem 8.3.4. Later on we shall use the same technique to obtain a cancellation law for exponentiation for co-ordinals.

8.3.5 THEOREM. If $A \neq 0$ and A, B, C are co-ordinals, then

$$AB = AC \Rightarrow B = C\,.$$

PROOF. Let $\mathsf{A} \in A$, $\mathsf{B} \in B$ and $\mathsf{C} \in C$ and suppose

$$p\colon \mathsf{AB} \simeq \mathsf{AC}\,.$$

Then $\mathsf{AB} \sim \mathsf{AC}$ and since AB and AC are well-orderings it follows from the preceding theorem that p is an extension of the unique minimal isomorphism, p_c, say, between AB and AC.

Now, classically,

$$\alpha \neq 0 \;\&\; \alpha\beta = \alpha\gamma \Rightarrow \beta = \gamma\,,$$

hence there is an isomorphism q_c (not necessarily partial recursive) such that

$$q_c\colon \mathsf{B} \sim \mathsf{C}\,.$$

Now the map

$$r_c\colon j(a,\, b) \Rightarrow j(a,\, q_c(b)),$$

defined only on C‘AB is an isomorphism between AB and AC and therefore, by theorem 8.3.4, p is an extension of r_c.

Since $A \neq 0$, there is an element, say a_0, in C‘A. Let p_0 be the map p with domain and range restricted to

$$\{j(a_0,\, n)\colon n \in \mathcal{N}\}\,,$$

then p_0 is partial recursive. Further, if $p_0 j(a_0, x)$ is defined, then its value is $j(a_0, y)$ for some y.

Now let q_0 be the map

$$q_0 : x \to l p_0 j(a_0, x),$$

then clearly q_0 is partial recursive and agrees with q_c on C'B (again by theorem 8.3.4). q_0 is one-one, since

$$q_0(x) = q_0(y) \Rightarrow l p_0 j(a_0, x) = l p_0 j(a_0, y)$$
$$\Rightarrow p_0 j(a_0, x) = j(a_0, c) \ \& \ p_0 j(a_0, y) = j(a_0, c)$$

(since $\rho p_0 \subseteq \{j(a_0, n) : n \in \mathcal{N}\}$ by construction)

$$\Rightarrow j(a_0, x) = j(a_0, y) \quad \text{(since } p \text{ is one-one)}$$
$$\Rightarrow x = y.$$

Thus q_0 is partial recursive and one-one and also agrees with q_c on C'B, i.e.

$$q_0 : \mathsf{B} \simeq \mathsf{C}$$

from which the theorem follows.

We have a stronger version of the above theorem which derives from SIERPINSKI (1948), however, we shall leave the proof of this theorem until appendix A.

8.3.6 LEMMA. If M is a principal number for multiplication, and $B, C \neq 0$, then

$$BC < M \Leftrightarrow B < M \ \& \ C < M.$$

PROOF. Suppose $BC < M$, then $B < M$ or $C = 1$ by theorem 8.3.3.(ii). In the former case $BM = M$ and in the latter trivially, $C < M$. Now

$$BC < M \Rightarrow BCM = M,$$

and therefore by theorem 8.3.5,

$$CM = M.$$

Using theorem 8.3.3.(ii) it follows that

$$C < M.$$

Conversely,

$$C < M \Rightarrow CM = M \quad \text{and} \quad B < M \Rightarrow BM = M.$$

Hence

$$(BC)\,M = B(CM) = BM = M$$

and by theorem 8.3.3.(ii)

$$BC < M\,.$$

8.4.

8.4.1 THEOREM. If A, C are quords then (i) If $A \neq 0$, then

$$B < C \Rightarrow AB < AC\,,$$

(ii) $B \leq C \Rightarrow AB \leq AC$.

PROOF. Let $\mathsf{A} \in A$ and $\mathsf{C} \in C$, then there exists $\mathsf{B} \in B$ such that $\mathsf{B} < \mathsf{C}$ and therefore for some quasi-well-ordering $\mathsf{D} \neq \emptyset$ we have

$$\mathsf{C} = \mathsf{B} \dotplus \mathsf{D}\,.$$

Now the reader will readily verify that

$$\mathsf{AC} = \mathsf{A}(\mathsf{B} \dotplus \mathsf{D}) = \mathsf{AB} \dotplus \mathsf{AD}\,.$$

But $\mathsf{AD} \neq \emptyset$, since A, $\mathsf{D} \neq \emptyset$, hence

$$\mathsf{AB} < \mathsf{AC}$$

and the theorem follows by taking C.O.T.s.

(ii) follows at once from (i).

8.4.2 THEOREM. There exist co-ordinals A, B, $C \neq 0$ such that

$$A < B \quad \text{but} \quad AC \nleq BC\,.$$

PROOF. Let $A = 1$, $B = V$ and $C = W$, then

$$AC = W \quad \text{and} \quad BC = VW\,.$$

By theorem 8.3.3.(i),

$$V \leq VW\,.$$

Hence if

$$W \leq VW,$$

V and W are comparable by theorem 4.2.8 which contradicts corollary 5.2.7.

8.4.3 THEOREM. If there is a principal number for multiplication, M, such that $B, C<M$ (or equivalently $BC<M$) then

$$A < B \Rightarrow AC \leq BC .$$

PROOF. If B or $C=0$ there is nothing to prove. Similarly if $A=0$. Otherwise, by lemma 8.3.6,

$$AC < M \quad \text{and} \quad BC < M .$$

Hence, by theorem 4.2.8, AC and BC are comparable. Now classically, for arbitrary ordinals α, β, γ

$$\alpha < \beta \Rightarrow \alpha\gamma \leq \beta\gamma ,$$

hence $AC \leq BC$ by corollary 5.3.5.

8.4.4 THEOREM. If A, B, C are co-ordinals, then

$$AC < BC \Rightarrow A < B .$$

PROOF. If $C=0$, then the assertion is trivial. If $C \neq 0$, then by theorem 8.3.3.(i),

$$A \leq AC \quad \text{and} \quad B \leq BC .$$

Hence by the transitivity of $\leq$ and theorem 4.2.8, A and B are comparable. By the classical theorem

$$\alpha\gamma < \beta\gamma \Rightarrow \alpha < \beta ,$$

we have

$$|A| < |B|$$

and hence by corollary 5.3.5

$$A < B .$$

8.4.5 THEOREM. There exist co-ordinals A, B, C such that

$$0 < AC \leq BC \quad \text{but} \quad A \nleq B .$$

PROOF. (As in the classical case.) Let $A=2$, $B=1$, $C=W$.

8.4.6 THEOREM. If B, C are comparable quords, then

$$AB < AC \Rightarrow B < C .$$

PROOF. Immediate from theorem 8.4.1.(i).

8.4.7 THEOREM. If A, B, C are co-ordinals and $AB<AC$ then $B<C$; similarly with "$\leq$" replacing "$<$" at both occurrences.

PROOF. $AB<AC \Rightarrow \alpha\beta<\alpha\gamma$ so by the classical result, $\beta<\gamma$. Now by corollary 5.3.6

$$(\mathrm{E!}D)\,(\mathrm{E!}E)\,(D+E=C\ \&\ |D|=\beta). \qquad (*)$$

So $AC=AD+AE$ by theorem 6.2.1. By theorem 4.2.1 we have $(\exists F)(A\cdot B+F=A\cdot C)$ and again using corollary 5.3.6 $(\mathrm{E!}G)\,(\mathrm{E!}H)(G+H=AC\ \&\ |G|=\alpha\beta)$.

We must have then $G=AB=AD$, and by theorem 8.3.5, $B=D$. $D<C$ by $(*)$ so the result follows.

The other part of the theorem follows from the above and theorem 8.3.5.

It is well-known that, classically, for arbitrary order types σ, τ,

$$\sigma\cdot n=\tau\cdot n\ \&\ n>0 \Rightarrow \sigma=\tau$$

(and similarly with "$\leq$" replacing "$=$" at both occurrences, see SIERPINSKI, (1948) for proofs). For quords the analogues are also true provided we use the strong interpretation of $\leq$; thus we have:

8.4.8 THEOREM. If A or B is a quord, then

$$A\cdot n+C=B\cdot n\ \&\ n>0 \Rightarrow (\exists D)\,(A+D=B).$$

PROOF[81] (by induction on n). If $n=1$, then the assertion is trivial. Now assume the theorem holds for n and that

$$A\cdot(n+1)+C=B\cdot(n+1).$$

Then

$$A\cdot n+(A+C)=B\cdot n+B$$

and by the directed refinement theorem 2.3.2, there is an E such that either

$$A\cdot n+E=B\cdot n \quad (\&\ A+C=E+B),$$

or

$$A\cdot n=B\cdot n+E\ \&\ E+A+C=B.$$

In the former case the assertion follows by the induction hypothesis. In the latter, by the induction hypothesis we have, for some F,

$$A=B+F,$$

whence

$$E + B + F + C = B$$

and by theorem 3.2.7,

$$E + B = B, \quad F = C = 0.$$

We conclude $A = B$.

8.5 We now introduce principal numbers for exponentiation and investigate the arithmetic laws for exponentiation.

8.5.1 DEFINITION. A co-ordinal (quord) A is said to be a *principal* (*$\mathcal{Q}$-principal*) *number for exponentiation* if $A > 2$ and

$$1 < B < A \Rightarrow B^A = A.$$

We write $\mathcal{H}$(exp) [$\mathcal{Q}$(exp)] for the collection of all principal ($\mathcal{Q}$-principal) numbers for exponentiation.

Later on we shall obtain an explicit description of all principal numbers for exponentiation (chapter 11).

8.5.2 THEOREM. If $A \in \mathcal{H}$(exp), then $|A|$ is a limit number.

PROOF. Left to the reader.

The condition in definition 8.5.1 is stronger than the condition

$$2 \le B, C < A \Rightarrow B^C < A. \tag{6}$$

This will be shown later by proving (lemma 11.2.2) that

$$2^A = A \Rightarrow W \text{ divides } A, \tag{7}$$

whereas V satisfies (6) but not (7).

8.5.3 THEOREM. If $A > 1$, B, C are co-ordinals, then

$$A^B = A^C \Rightarrow B = C.$$

PROOF.[82] Let $\mathsf{A} \in A$, $\mathsf{B} \in B$ and $\mathsf{C} \in C$ and suppose

$$p: \mathsf{A}^{\mathsf{B}} \simeq \mathsf{A}^{\mathsf{C}}.$$

Then $\mathsf{A}^{\mathsf{B}} \sim \mathsf{A}^{\mathsf{C}}$ and since A^{B} and A^{C} are well-orderings it follows from theorem 8.3.4 that p is an extension of the unique minimal isomorphism p_c, say, between A^{B} and A^{C}. Now, classically,

$$\alpha > 1 \;\&\; \alpha^{\beta} = \alpha^{\gamma} \to \beta = \gamma ,$$

hence there is an isomorphism q_c (not necessarily partial recursive) such that

$$q_c : \mathsf{B} \sim \mathsf{C} .$$

Now the map

$$r_c : e\begin{pmatrix} b_0 \dots b_n \\ a_0 \dots a_n \end{pmatrix} \to e\begin{pmatrix} q_c(b_0) \dots q_c(b_n) \\ a_0 \quad \dots \quad a_n \end{pmatrix}$$

defined only on[83] $e(\mathsf{A}, \mathsf{B})$ is an isomorphism between A^{B} and A^{C}. Hence by theorem 8.3.4, p is an extension of r_c.

Since $A > 1$, there is a non-minimum element, say a_0, in C‘A. Let p_0 be the map p with domain and range restricted to

$$\left\{ e\begin{pmatrix} n \\ a_0 \end{pmatrix} : n \in \mathcal{N} \right\},$$

then p_0 is partial recursive. Further if

$$p_0 \, e\begin{pmatrix} x \\ a_0 \end{pmatrix}$$

is defined then its value is

$$e\begin{pmatrix} y \\ a_0 \end{pmatrix}$$

for some y. Now let q_0 be the map

$$q_0 : x \to l\left(\left(p_0 \, e\begin{pmatrix} x \\ a_0 \end{pmatrix}\right)_0\right),^{84}$$

then clearly q_0 is partial recursive and agrees with q_c on C‘B (again by theorem 8.3.4). q_0 is one-one, since

$$\begin{aligned} q_0(x) = q_0(y) &\Rightarrow l\left(\left(p_0 \, e\begin{pmatrix} x \\ a_0 \end{pmatrix}\right)_0\right) = l\left(\left(p_0 \, e\begin{pmatrix} y \\ a_0 \end{pmatrix}\right)_0\right) \\ &\Rightarrow p_0\left(e\begin{pmatrix} x \\ a_0 \end{pmatrix}\right) = 2^{j(u,\, q_0(x))} \cdot 3^{x_1} \cdot \dots \cdot p_n^{x_n} \\ &\;\&\; p_0\left(e\begin{pmatrix} y \\ a_0 \end{pmatrix}\right) = 2^{j(v,\, q_0(y))} \cdot 3^{y_1} \cdot \dots \cdot p_m^{y_m} \end{aligned}$$

for some $u, v, m, n, x_1, \ldots, x_n, y_1, \ldots, y_m$. But by the definition of p_0, any image of p_0 is of the form

$$2^{j(a_0,\, b)}$$

and hence $n=m=0$, $u=v=a_0$ and

$$p_0\left(e\begin{pmatrix} x \\ a_0 \end{pmatrix}\right) = 2^{j(a_0,\, q_0(x))}$$

and

$$p_0\left(e\begin{pmatrix} y \\ a_0 \end{pmatrix}\right) = 2^{j(a_0,\, q_0(y))}.$$

Therefore, from $q_0(x)=q_0(y)$, we have

$$p_0\left(e\begin{pmatrix} x \\ a_0 \end{pmatrix}\right) = p_0\left(e\begin{pmatrix} y \\ a_0 \end{pmatrix}\right),$$

from which it follows, since p_0 and $\boldsymbol{e}$ are one-one, that $x=y$. Thus we have shown that q_0 is a recursive isomorphism between B and C and the proof is complete.

8.5.4 THEOREM. There exist co-ordinals A, B, C, all >1, such that

$$A^C = B^C \quad \text{but} \quad A \neq B.$$

PROOF (as in the classical case). Let $A=2$, $B=3$, $C=W$, then by theorem 7.4.2, $2^W=3^W=W$.

8.5.5 THEOREM. (i) $A>1 \ \& \ B<C \Rightarrow A^B \leq A^C$ (for arbitrary quords A, B, C).

(ii) If C is a co-ordinal, then $A>1 \ \& \ B<C \Rightarrow A^B < A^C$.

PROOF. (i) Let $\mathsf{A} \in A$ and $\mathsf{C} \in C$, then there exists $\mathsf{B} \in B$ such that $\mathsf{B} < \mathsf{C}$ and therefore, for some quasi-well-ordering D

$$\mathsf{C} = \mathsf{B} \mathbin{\hat{+}} \mathsf{D}.$$

Now, if $d \in \mathrm{C}`\mathsf{D}$, then for all $b \in \mathrm{C}`\mathsf{B}$ we have

$$d >_{\mathsf{B} \hat{+} \mathsf{D}} b,$$

hence any bracket symbol in $(\mathsf{A}, \mathsf{B} \mathbin{\hat{+}} \mathsf{D})$ is of the form

$$\begin{pmatrix} d_0 \ldots d_m \, b_0 \ldots b_n \\ a_0 \ldots a_m \, a_0' \ldots a_n' \end{pmatrix},$$

where the $d_i \in C`D$, $b_j \in C`B$ and the $a_i, a'_j \in C`A$ $(i=0,\ldots,m; j=0,\ldots,n)$. Hence from the definition of exponentiation we easily obtain

$$A^B \leq A^{B\hat{+}D} = A^C.$$

(ii) By theorem 7.2.1.(ii), assuming $A \in A$, etc. as above, A^C (and hence A^B) is a quasi-well-ordering. Hence it suffices, by lemma 4.2.5, to prove that

$$A^B \neq A^C.$$

Now $A > 1$, hence there exists $a \in C`A$ with $a \neq \min(A)$. Since $D \neq \emptyset$ there also exists $d \in C`D$. Therefore

$$e\binom{d}{a} \in C`A^C - C`A^B,$$

from which we have the required result.

8.5.6 LEMMA. (i) If C is a co-ordinal and $A, C > 1$ then $A < A^C$.

(ii) If $C \geq 1$, then $A \leq A^C$ (for arbitrary quords A, C).

PROOF. (i) follows at once from the preceding theorem.

(ii) $C \geq 1 \Rightarrow (\exists D)\,(C = 1 + D)$. Therefore

$$A^C = A^{1+D} = A \cdot A^D \quad \text{by theorem 7.3.4}.$$

Now, by the definition of exponentiation we have at once

$$1 \leq A^D \quad \text{since} \quad A \neq 0.$$

Hence

$$\begin{aligned} A^C &= A(1+E) \quad \text{for some } E, \\ &= A + AE \quad \text{by theorem 6.2.1}. \end{aligned}$$

The required result now follows at once.

8.5.7 THEOREM. There exist co-ordinals A, B, C such that

$$A < B \quad \text{but} \quad A^C \not\leq B^C.$$

PROOF. Let $A = 2$, $B = V$ and $C = W$, then by theorem 7.4.2,

$$A^C = C = W$$

and by lemma 8.5.6.(i),

$$V < V^W = B^C.$$

Now, if

$$A^C \leq B^C,$$

then by theorem 4.2.8, V and W are comparable which contradicts corollary 5.2.7.

Thus we see that the analogue of one of the classical laws for exponentiation breaks down in a very similar way to one of the multiplicative laws (theorem 8.4.2). However, the similarity also extends to the cases where the analogues do go over.

8.5.8 LEMMA. If E is a principal number for exponentiation, then $A, B<E \Rightarrow A^B<E$ and conversely if $A, B>1$.

PROOF. The assertion is trivial if $A, B \leq 1$. Otherwise, if E is a principal number for exponentiation, then $A<E \Rightarrow A^E=E$ and similarly for B; moreover, we must also have that A and B are co-ordinals since E is a co-ordinal. It follows that

$$A^{(BE)} = E.$$

Since $B<E$ it follows by theorem 2.1.2 that there is a (co-ordinal) C such that

$$B + C = E.$$

Therefore

$$E = A^{(BE)} = A^{(B+C)} = A^B \cdot A^C.$$

But

$$A^C > 1, \quad \text{since} \quad C \neq 0;$$

hence

$$A^C = 1 + D \quad \text{for some} \quad D \neq 0$$

and it follows that

$$E = A^B(1 + D) = A^B + A^B D \quad \text{where} \quad A^B D \neq 0,$$

i.e.

$$A^B < E.$$

Conversely, suppose

$$A, B > 1 \quad \text{and} \quad A^B < E,$$

then by lemma 8.5.6.(i),

$$A < E.$$

Since E is a principal number for exponentiation,

$$E = A^E = (A^B)^E = A^{BE}.$$

By theorem 8.5.3 it follows that

$$BE = E$$

and hence by theorem 8.3.3.(ii) we have

$$B < E.$$

8.5.9 THEOREM. If there is a principal number for exponentiation, E, such that $B, C < E$, then

$$A < B \Rightarrow A^C \leq B^C.$$

PROOF. By the transitivity of $\leq$ and lemma 8.5.8,

$$A^C < E \quad \text{and} \quad B^C < E.$$

Hence by theorem 4.2.8, A^C and B^C are comparable co-ordinals. Now, classically, for ordinals α, β, γ,

$$\alpha < \beta \Rightarrow \alpha^\gamma \leq \beta^\gamma;$$

hence

$$A < B \Rightarrow A^C \leq B^C.$$

8.5.10 THEOREM. If A, B, C are co-ordinals, then

$$A^C < B^C \Rightarrow A < B.$$

PROOF. If $C=0$, there is nothing to prove. Otherwise, by lemma 8.5.6(ii),

$$A \leq A^C \quad \text{and} \quad B \leq B^C$$

and therefore by theorem 4.2.8 and the transitivity of $\leq$, A and B are comparable. Hence by the classical theorem for ordinals

$$\alpha^\gamma < \beta^\gamma \Rightarrow \alpha < \beta,$$

we have

$$|A| < |B| \quad \text{and hence} \quad A < B.$$

8.5.11 THEOREM. There exist co-ordinals A, B, C such that

$$1 < A^C \leq B^C \quad \text{but} \quad A \nleq B.$$

PROOF (as in the classical case). Let $A=3$, $B=2$ and $C=W$, then by theorem 7.4.2,

$$A^C = B^C = W.$$

8.5.12 THEOREM. (i) If B, C are comparable and $A>1$, then

$$A^B < A^C \Rightarrow B < C.$$

PROOF. By theorem 8.5.5., since $C \leq B$ means $C=B$ or $C<B$.

CHAPTER 9

CANTOR NORMAL FORMS

9.1 In this chapter we prove the existence of Cantor Normal Forms[91] for a large class of co-ordinals, but first of all we prove a general decomposition theorem for all co-ordinals less than some principal number for addition.

We restrict our attention from now until the end of Part One of this monograph to co-ordinals, returning to the consideration of quords in Part Two. Since we are dealing with co-ordinals we shall repeatedly use the fact (theorem 4.2.2) that

$$A < B \Leftrightarrow (\exists C)\,(A + C = B \;\&\; C \neq 0).$$

9.1.1 LEMMA. A co-ordinal P is a principal number for addition if, and only if,

$$A < P \Rightarrow A \cdot W \leq P .$$

PROOF. Immediate from lemma 6.3.1 and definition 8.1.3.

9.1.2 THEOREM. If $P \in \mathscr{H}(+)$, then $P \cdot W \in \mathscr{H}(+)$ and there is no principal number Q such that

$$P < Q < P \cdot W .$$

PROOF. Suppose $P \in \mathscr{H}(+)$ and $A < P \cdot W$, then by theorem 6.2.5, there exist n, D such that

$$A = P \cdot n + D , \quad \text{where} \quad D < P .$$

Hence

$$\begin{aligned} A + P\cdot W &= A + (P + P\cdot W) && \text{since } 1 + W = W, \\ &= (A + P) + P\cdot W \\ &= (P\cdot n + D + P) + P\cdot W \\ &= P\cdot(n+1) + P\cdot W && \text{since } P \in \mathscr{H}(+) \text{ and } D < P, \\ &= P\cdot W. \end{aligned}$$

Hence $P\cdot W \in \mathscr{H}(+)$. Clearly

$$P < P\cdot W$$

since $P \neq 0$.

Now suppose $Q \in \mathscr{H}(+)$ and $P < Q$, then by lemma 9.1.1,

$$P\cdot W \leq Q;$$

hence

$$Q \not< P\cdot W.$$

9.1.3 THEOREM.[92] If $0 < A < P \in \mathscr{H}(+)$, then there exist principal numbers for addition $P_1, \ldots, P_n$ such that

$$P > P_n \geq \cdots \geq P_1 \tag{1}$$

and

$$A = P_n + \cdots + P_1. \tag{2}$$

Further, if

$$A = Q_m + \cdots + Q_1$$

is any other representation of A as a sum of principal numbers Q_i $(i = 1, \ldots, m)$ such that

$$Q_m \geq \cdots \geq Q_1,$$

then

$$m = n \;\&\; P_i = Q_i \quad \text{for} \quad i = 1, \ldots, n.$$

Conversely, if A is expressible in the form (2) with $P_n \geq \cdots \geq P_1$ where the P_i $(i = 1, \ldots, n)$ are principal numbers for addition, then there is a principal number P e.g. $P_n\cdot W$, such that

$$P > A \quad \text{and} \quad P > P_i \quad \text{for } i = 1, \ldots, n.$$

PROOF (by transfinite induction on the partial well-ordering $\leq$). We assume $0 < A < P \in \mathscr{H}(+)$ and take as induction hypothesis: If $0 < B < A$

then B is uniquely expressible in the form (2) where the P_i are principal numbers satisfying (1).

If A is a principal number for addition then the theorem follows at once from theorem 9.1.2. Now suppose A is not a principal number, then there exist B, C such that

$$B + C = A, C \neq 0, A \quad \text{and} \quad B + C < P. \tag{3}$$

By corollary 8.1.6,

$$C < A.$$

Let P_1 be the least C satisfying (3). P_1 exists since

$$\{C : C < A\}$$

is well-ordered by $\leq$. We show that P_1 is a principal number for addition. Suppose

$$P_1 = D + E,$$

then by corollary 8.1.6,

$$E < P$$

and hence by theorem 4.2.8, P_1 and E are comparable. But

$$|E| \leq |P_1|,$$

hence

$$E \leq P_1$$

and by the minimality of P_1,

$$E = P_1.$$

Therefore P_1 is a principal number by theorem 8.1.2. Now let B_1 be the least B such that

$$B + P_1 = A.$$

If $B_1 = 0$, then A is a principal number which contradicts our assumption. Hence

$$B_1 \neq 0$$

and by the induction hypothesis B_1 has a unique decomposition

$$B = P_n + \cdots + P_2,$$

where

$$P > P_n \geq \cdots \geq P_2$$

and all the P_i $(i=2,\ldots,n)$ are principal numbers. Therefore

$$A = P_n + \cdots + P_1$$

and, since

$$P_1 < P,$$

all the P_i $(i=1,\ldots,n)$ are comparable by theorem 4.2.8. Suppose

$$P_2 < P_1,$$

then, since P_1 is a principal number for addition,

$$P_2 + P_1 = P_1;$$

hence

$$A = (P_n + \cdots + P_2) + P_1 = (P_n + \cdots + P_3) + P_1 .$$

Now

$$P_2 \neq 0 \Rightarrow P_n + \cdots + P_3 < P_n + \cdots + P_2 = B_1 ,$$

whereas B_1 was chosen as the least B such that $B + P_1 = A$. This contradiction shows that

$$P_2 \geq P_1$$

and we conclude that A has a decomposition of the form (2).

As regards uniqueness, let

$$A = P_n + \cdots + P_1 \quad \text{and} \quad A = Q_m + \cdots + Q_1$$

be two decompositions of A as a sum of non-increasing principal numbers. By theorem 4.2.8, P_n and Q_m are comparable. Suppose

$$P_n > Q_m,$$

then, since P_n is a principal number for addition,

$$Q_m + P_n = P_n .$$

Therefore

$$\begin{aligned} A &= Q_m + P_n + \cdots + P_1 \\ &= Q_m + Q_m + P_n + \cdots + P_1 \\ &= \ldots \\ &= Q_m \cdot (m+1) + A . \end{aligned}$$

Since $A \neq 0$, we therefore have

$$Q_m \cdot (m+1) < A . \tag{4}$$

On the other hand, if $i \leq m$, then

$$Q_i + Q_m = Q_m \quad \text{or} \quad Q_m \cdot 2,$$

according as

$$Q_i < Q_m \quad \text{or} \quad Q_i = Q_m.$$

Consequently, we have

$$Q_m \cdot (m+1) + A = A < A + Q_m \cdot m \leq Q_m \cdot 2m$$

and by corollary 3.2.9, it follows that

$$A \leq Q_m \cdot m, \tag{5}$$

which contradicts (4) above. We conclude

$$Q_m \nless P_n \quad \text{and similarly} \quad P_n \nless Q_m,$$

but since P_n and Q_m are comparable it follows that

$$P_n = Q_m \quad \text{and} \quad P_{n-1} + \cdots + P_1 = Q_{m-1} + \cdots + Q_1$$

(by corollary 3.2.9). Proceeding thus we obtain after s(= minimum of m, n) steps

$$P_{n-r} = Q_{m-r} \quad r = 0, \ldots, s$$

and either

$$P_t + \cdots + P_1 = 0 \quad \text{or} \quad Q_t + \cdots + Q_1 = 0,$$

where t = maximum of $m-s, n-s$. By theorem 2.2.4.(ii) it follows that $t=0$ and hence that $m=n$ and

$$P_i = Q_i \quad \text{for} \quad i = 1, \ldots, n(=m).$$

For the converse: if

$$A = P_n + \cdots + P_1,$$

then by the argument which produced (5) we have

$$A \leq P_n \cdot n$$

and consequently

$$A < P_n \cdot W.$$

By theorem 9.1.2, $P_n \cdot W$ is a principal number for addition. This completes the proof.

9.2.

9.2.1 LEMMA. $C<A+B \Leftrightarrow [C<A \vee (\exists D)(D<B \;\&\; C=A+D)]$.

PROOF. Immediate from the directed refinement theorem 2.3.2.

The next lemma is a strengthening of theorem 6.2.5.

9.2.2 LEMMA. If $A \neq 0$, then

$$C < AB \Leftrightarrow (\exists D)\,(D < B \;\&\; AD \leq C < A(D+1)).$$

PROOF. The implication from right to left is clear from corollary 5.2.3, theorem 8.4.1 and the transitivity of $<$ for co-ordinals.

Suppose now that

$$C < AB,$$

then there exists $F \neq 0$ such that

$$C + F = AB.$$

Given $\mathsf{A} \in A$ and $\mathsf{B} \in B$, then by the separation lemma 2.3.1, there exists $\mathsf{C} \in C$ and $\mathsf{F} \in F$ such that

$$[\mathsf{C})\,(\mathsf{F} \text{ and}] \quad \mathsf{C} \mathbin{\hat{+}} \mathsf{F} = \mathsf{A} \cdot \mathsf{B}.$$

Let

$$f = \min(\mathsf{F}),\ d = l(f),\ \mathsf{D} = d)\,\mathsf{B} \quad \text{and} \quad D = \mathrm{COT}(\mathsf{D}).$$

We shall show that for this D,

$$AD \leq C < A(D+1).$$

Now

$$\mathsf{AD} \subseteq \mathsf{C} \subseteq \mathsf{AB},$$

since

$$\begin{aligned} j(a, x) \in \mathrm{C}`\mathsf{AD} &\Rightarrow x \leq_{\mathsf{B}} d \\ &\Rightarrow j(a, x) \leq_{\mathsf{AB}} f \\ &\Rightarrow j(a, x) \in \mathrm{C}`\mathsf{C}. \end{aligned}$$

Further, since AD and C are both initial segments of AB, we must have

$$\mathsf{AD} \leq \mathsf{C}.$$

By corollary 5.2.3,

$$D < B \Rightarrow D + 1 \leq B$$

and hence by theorem 8.4.1.(ii) it follows that

$$A(D+1) \leq AB.$$

From theorem 4.2.8 we now have

$$A(D+1) \leq C \quad \text{or} \quad C \leq A(D+1),$$

but

$$\mathsf{C} \subseteq \mathsf{A} \cdot (\mathsf{D} \mathbin{\hat{+}} \{\langle d, d \rangle\})$$

and so we have

$$C \leq A(D+1).$$

Finally, since

$$f \in \mathrm{C}`j(\mathsf{A}, d) - \mathrm{C}`\mathsf{C}, \quad C \neq A(D+1)$$

and the proof is complete.

9.2.3 Lemma. $1 \leq C < A^B \Leftrightarrow (\exists D)\,(D < B \;\&\; A^D \leq C < A^{D+1})$.

Proof. The implication from right to left follows at once from corollary 5.2.3, theorem 8.5.5.(ii) and the transitivity of $<$. Now suppose

$$1 \leq C < A^B,$$

then there exists $F \neq 0$ such that

$$C + F = A^B.$$

Further, there exist A, B such that

$$\mathsf{A} = [\mathrm{A}] \in A \quad \text{and} \quad \mathsf{B} = [\mathrm{B}] \in B$$

and by the separation lemma 2.3.1, there exist $\mathsf{C} \in C$ and $\mathsf{F} \in F$ such that

$$\mathsf{C} \mathbin{\hat{+}} \mathsf{F} = \mathsf{A}^{\mathsf{B}}.$$

Let

$$f = \min(\mathsf{F}),$$

then

$$f = e\begin{pmatrix} b_0 \ldots b_n \\ a_0 \ldots a_n \end{pmatrix},$$

where $b_0 \succ \ldots \succ b_n$, $b_i \in \mathrm{B}$ $(i = 0, \ldots, n)$, $n \neq -1$ (since $C \geq 1$) and 93
$0 \neq a_i \in \mathrm{A}$ $(i = 0, \ldots, n)$.

Also let

$$\mathsf{D} = b_0)\mathsf{B} \quad \text{and} \quad D = \mathrm{COT}(\mathsf{D}),$$

then we claim

$$\mathsf{A}^{\mathsf{D}} \leq \mathsf{C}.$$

Now $x \in \mathrm{C}`\mathsf{A}^{\mathsf{D}}$ implies

$$x = e\begin{pmatrix} d_0 \dots d_n \\ a'_0 \dots a'_n \end{pmatrix},$$

where $d_0 \prec b_0$ and it follows that

$$x <_{\mathsf{A}\ \exp\ \mathsf{B}} f,$$

which implies $x \in \mathrm{C}`\mathsf{C}$. Further, if $y \in \mathrm{C}`\mathsf{A}^{\mathsf{D}}$ and $\langle y, x\rangle \in \mathsf{A}^{\mathsf{D}}$, then by the same argument

$$y <_{\mathsf{A}\ \exp\ \mathsf{B}} f.$$

It follows at once (from corollary 4.2.5) that

$$A^D \leq C.$$

Now let

$$\mathsf{E} = \mathsf{D} \hat{+} \{\langle b_0, b_0\rangle\},$$

then

$$x \in \mathrm{C}`\mathsf{C} \Rightarrow x = 0 \quad \text{or} \quad x = e\begin{pmatrix} b'_0 \dots b'_k \\ a''_0 \dots a''_k \end{pmatrix},$$

where $b'_0 \succ \dots \succ b'_k$, $b'_i \in \mathrm{B}$ and $0 \neq a''_i \in \mathrm{A}$. Further, by the definition of b_0,

$$\langle b'_0, b_0\rangle \in \mathsf{B}$$

and hence $x \in \mathrm{C}`\mathsf{A}^{\mathsf{E}}$ and

$$\mathsf{C} \leq \mathsf{A}^{\mathsf{E}} \in A^{D+1}.$$

But $\mathsf{C} \neq \mathsf{A}^{\mathsf{E}}$ since

$$f \in \mathrm{C}`\mathsf{A}^{\mathsf{E}} - \mathrm{C}`\mathsf{C}$$

and it follows that

$$A^D \leq C < A^{D+1}.$$

9.2.4 LEMMA. $A \geq W \ \&\ B < C \Rightarrow A^B + A^C = A^C$.

PROOF. Since $B < C$, $(\exists D)\,(D \neq 0 \ \&\ B + D = C)$.

Hence

$$\begin{aligned} A^B + A^C &= A^B + A^B \cdot A^D && \text{by theorem 7.3.4}, \\ &= A^B(1 + A^D) && \text{by theorem 6.2.1}, \\ &= A^B(1 + A + E) && \text{by lemma 8.5.6.(i)}, \\ & && \quad \text{where } A + E = A^D, \end{aligned}$$

$$
\begin{aligned}
&= A^B(1 + W + F + E) && \text{since } W \leq A \Rightarrow W + F = A \\
& && \text{for some } F\,, \\
&= A^B(W + F + E) && \text{by theorem 5.2.6.(i)}\,, \\
&= A^B A^D = A^C\,.
\end{aligned}
$$

9.2.5 LEMMA. If $C<AB$, then there exist unique Q, R such that

$$C = AQ + R\,, \quad \text{where} \quad 0 \leq Q < B \;\&\; 0 \leq R < A\,.$$

PROOF. By lemma 9.2.2,

$$C < AB \Rightarrow (\exists Q)\,(0 \leq Q < B \;\&\; AQ \leq C < A(Q + 1))\,.$$

Hence

$$AQ \leq C < AQ + A\,.$$

Since A, C, Q are co-ordinals,

$$R = C - AQ$$

is well-defined, $0 \leq R < A$ and $C = AQ + R$ as required. Now suppose

$$C = AQ + R = AQ_1 + R_1\,,$$

where $0 \leq Q_1 < B$ and $0 \leq R_1 < A$. Then, since we also have

$$0 \leq Q < B \quad \text{and} \quad 0 \leq R < A\,,$$

by theorem 4.2.8 Q and Q_1 are comparable (and so too are R and R_1). By corollary 5.2.3, if $Q \neq Q_1$ then either

$$Q_1 + 1 \leq Q \quad \text{or} \quad Q + 1 \leq Q_1\,.$$

Suppose the former holds, then

$$AQ_1 \leq C < A\,(Q_1 + 1) \leq AQ < C\,,$$

which is a contradiction. Similarly we cannot have $Q+1 \leq Q_1$, hence

$$Q = Q_1$$

and by corollary 3.2.9,

$$R = R_1\,.$$

9.2.6 LEMMA. If $1 \leq C < A^B$ then there exist unique D, Q, R such that

$$C = A^D Q + R\,, \quad \text{where} \quad D < B, \quad 0 < Q < A \quad \text{and} \quad 0 \leq R < A^D\,.$$

PROOF. By lemma 9.2.3,

$$1 \leq C < A^B \Rightarrow (\exists D)\,(A^D \leq C < A^{D+1} \;\&\; D < B),$$

hence

$$C < A^{D+1} = A^D \cdot A .$$

Now, if $A < C$, $D \neq 0$. By lemma 9.2.5, there therefore exist Q, R such that

$$C = A^D Q + R, \quad \text{where} \quad 0 < Q < A \quad \text{and} \quad 0 \leq R < A .$$

Suppose that we also had

$$C = A^E Q_1 + R_1, \quad \text{where} \quad E < B, \quad 0 < Q_1 < A \quad \text{and} \quad 0 \leq R_1 < A .$$

By theorem 4.2.8, D and E are comparable and hence by corollary 5.2.3, if $D \neq E$, then

$$D + 1 \leq E \quad \text{or} \quad E + 1 \leq D .$$

Suppose the former holds, then by theorem 8.5.5.(i), and using the fact that A^E is a co-ordinal, we have

$$A^D \leq C < A^{D+1} \leq A^E \leq C,$$

which is a contradiction. Likewise we cannot have $E+1 \leq D$ and we conclude

$$D = E .$$

Now suppose $A \geq C$, then either trivially $A = C$ or

$$C = A^0 Q + R, \quad \text{where} \quad 0 < Q < A \quad \text{and} \quad R = 0$$

and Q, R are unique by the preceding lemma.

Thus for all A satisfying the hypothesis we obtain D, Q, R uniquely as required.

9.2.7 THEOREM. If $1 \leq C < A^B$ then C is uniquely expressible in the form

$$C = A^{B_1} \cdot D_1 + \cdots + A^{B_r} \cdot D_r, \tag{1}$$

where $B > B_1 > \cdots > B_r$ and $1 \leq D_i < A$ for $1 \leq i \leq r$.

PROOF. By lemma 9.2.6,

$$1 \leq C < A^B \Rightarrow C = A^{B_1} \cdot D_1 + C_1,$$

where $0 \leq D_1 < A$, $0 \leq C_1 < A^{B_1}$ and $B_1 < B$ and B_1, D_1, C_1 are unique. Since

$$C_1 < A^{B_1} < A^B \quad \text{and} \quad A^{B_1} \leq C < A^B$$

it follows that

$$C_1 < C < A^B.$$

Since $\{E: E < A^B\}$ is well-ordered by $\leq$ we obtain, by a finite number of repetitions of the above process, the decomposition (1). Suppose that C is also expressible as

$$C = A^{F_1} \cdot H_1 + \cdots + A^{F_s} \cdot H_s,$$

where $B > F_1 > \cdots > F_s$ and $1 \leq H_i < A$ for $1 \leq i \leq s$; then by theorem 4.2.8 and theorem 8.3.3(i), A^{B_1} and A^{F_1} are comparable. If $A^{B_1} < A^{F_1}$ then $B_1 + 1 \leq F_1$ and

$$A^{B_1} \leq C < A^{B_1+1} \leq A^{F_1} \leq C,$$

which is a contradiction. Similarly $A^{F_1} \nless A^{B_1}$, whence $A^{B_1} = A^{F_1}$ and by theorem 8.5.3, $B_1 = F_1$. Hence

$$C = A^{B_1} \cdot D_1 + C_1 = A^{B_1} \cdot H_1 + E_1,$$

where $0 < D_1 < A$ and $0 < H_1 < A$. If $D_1 \neq H_1$, then by corollary 5.2.3,

$$D_1 + 1 \leq H_1 \quad \text{or} \quad H_1 \leq D_1 + 1.$$

Suppose the former holds, then

$$C = A^{B_1} \cdot D_1 + C_1 < A^{B_1}(D_1 + 1) \leq A^{B_1} \cdot H_1 \leq A^{B_1} \cdot H_1 + E_1 = C,$$

which is a contradiction. Similarly, $H_1 \nleq D_1 + 1$ and we conclude

$$D_1 = H_1.$$

By corollary 3.2.9 it follows that

$$C_1 = E_1$$

and hence by induction on the maximum of r, s that

$$B_i = F_i \quad \text{and} \quad D_i = H_i \quad \text{for} \quad 1 \leq i \leq r = s.$$

This completes the proof.

As an immediate consequence we obtain our main theorem of this section.

9.2.8 Theorem (Cantor normal form). If $0 < C < W^A$, then C is uniquely expressible in the form

$$C = W^{A_1} \cdot n_1 + \cdots + W^{A_r} \cdot n_r, \tag{2}$$

where $A > A_1 > \cdots > A_r$ and the n_i are finite, non-zero co-ordinals.

9.2.9 COROLLARY. A necessary and sufficient condition that a co-ordinal $C \neq 0$ should have a Cantor Normal Form (2) is that there should exist a co-ordinal A such that $C < W^A$.

PROOF. The condition is obviously sufficient. Now suppose C has the given Cantor Normal Form and let A be greater than A_1 (say, take $A = A_1 + 1$); then by $n_1 + \cdots + n_r$ applications of lemma 9.2.4, we obtain $C + W^A = W^A$ whence $C < W^A$.

CHAPTER 10

UNIQUENESS RESULTS

10.1 We showed earlier (theorems 4.1.4, 5.2.4) that the finite co-ordinals are unique but that there exist c mutually incomparable co-ordinals of classical ordinal ω. Shortly we shall show that this latter result holds for all infinite denumerable ordinals. However, we shall go on to give criteria for some collections of co-ordinals to contain precisely one representative for each member of a given collection of classical ordinals. In this chapter we shall give simple criteria for sufficiently small well-orderings to be 'natural' in the sense that if two well-orderings of the same (classical) type are natural then they are recursively isomorphic. Clearly, it is sufficient to show that the co-ordinal associated with such a natural well-ordering is uniquely determined. In fact, it will turn out that natural well-orderings are recursively enumerable. Here we shall only deal with addition and multiplication and shall treat exponentiation in chapter 11. The results can be extended further by considering additional functions (*see* e.g. CROSSLEY and PARIKH, 1963) but at present no uniform procedure is available for obtaining natural well-orderings of larger ordinals and indeed it is not known whether natural well-orderings exist for all the recursive ordinals (CHURCH and KLEENE, 1936).

10.1.1 DEFINITION.[101] If $\mathscr{A}$ is a collection of co-ordinals, then $\mathscr{A}$ is said to be *α-unique* if

$$A, B \in \mathscr{A} \;\&\; |A| = |B| < \alpha \Rightarrow A = B.$$

$\mathscr{A}$ is said to be *strictly α-unique* if $\mathscr{A}$ is α-unique but not β-unique for any $\beta > \alpha$.

10.1.2 COROLLARY. $\mathscr{C}$ is strictly ω-unique.

PROOF. Immediate from theorems 4.1.4, 5.2.4.

We have the following strengthening of this result.

10.1.3 THEOREM. If α is any infinite denumerable ordinal then there exist c mutually incomparable co-ordinals of ordinal α.

PROOF. If α is infinite, then $\alpha=\omega+\beta$ for some β. Let B be a fixed co-ordinal of ordinal β. (Such a co-ordinal exists since there exists a well-ordering of type β and hence a well-ordering embeddable in R; though, of course, the embedding may not be partial recursive.) Let V_1, V_2 be two incomparable co-ordinals of ordinal ω then by corollary 5.3.5, if V_1+B is comparable with V_2+B then $V_1+B=V_2+B$ and by theorem 4.2.8, V_1 and V_2 are comparable which contradicts the choice of V_1 and V_2.

Since, by theorem 5.2.4, there are c mutually incomparable co-ordinals of type ω, the theorem now follows at once.

We now proceed to prove that the collections of principal numbers for addition and multiplication are strictly ω^{ω}- and $\omega^{\omega^{\omega}}$-unique, respectively. In order to do this we prove certain closure conditions for such principal numbers and then we establish necessary conditions for "small" co-ordinals to be principal numbers.

10.1.4 THEOREM. If $P_1, P_2 \in \mathscr{H}(+)$, then

$$P_1 + P_2 \in \mathscr{H}(+) \Leftrightarrow P_1 < P_2 \Leftrightarrow P_1 + P_2 = P_2 .$$

PROOF. $P_1 < P_2 \Rightarrow P_1 + P_2 = P_2$ by the definition of $\mathscr{H}(+)$.

$$\begin{aligned} &P_1 + P_2 \in \mathscr{H}(+) \\ &\qquad \Rightarrow P_1 + (P_1 + P_2) = P_1 + P_2 \\ &\qquad \Rightarrow P_1 + P_2 = P_2 \quad \text{by corollary 3.2.9} \\ &\qquad \Rightarrow P_1 < P_2 . \end{aligned}$$

10.1.5 THEOREM. If $P_1, P_2 \in \mathscr{H}(+)$, then $P_1 \cdot P_2 \in \mathscr{H}(+)$.

PROOF. By lemma 9.2.5, $A < P_1 P_2$ implies

$$(\exists Q)\,(\exists R)\,(A = P_1 Q + R \;\&\; 0 \le Q < P_2 \;\&\; 0 \le R < P_1) .$$

Hence

$$
\begin{aligned}
A + P_1P_2 &= (P_1Q + R) + P_1P_2 \\
&= P_1Q + (R + P_1 + P_1P_2) && \text{since } P_2 \text{ is infinite and } 1 + P_2 = P_2 \\
& && \text{by definition of } \mathscr{H}(+), \\
&= P_1Q + (P_1 + P_1P_2) && \text{since } R < P_1, \\
&= P_1Q + P_1P_2 \\
&= P_1(Q + P_2) \\
&= P_1P_2 && \text{since } Q < P_2.
\end{aligned}
$$

10.1.6 THEOREM. If $P \in \mathscr{H}(+)$ and A is any co-ordinal > 0, then $P^A \in \mathscr{H}(+)$.

PROOF. By theorem 9.2.7, $0 < C < P^A$ implies

$$C = P^{A_1} \cdot D_1 + \cdots + P^{A_r} \cdot D_r,$$

where $A > A_1 > \cdots > A_r$ and $0 < D_i < P$ for $1 \leq i \leq r$.

$$
\begin{aligned}
P^{A_r} \cdot D_r + P^A &= P^{A_r}(D_r + P^E) && \text{where } A = A_r + E, \\
&= P^{A_r}(D_r + P + F) && \text{where } P + F = P^E, \\
&= P^{A_r}(P + F) && \text{since } P \in \mathscr{H}(+), \\
&= P^{A_r} \cdot P^E \\
&= P^A.
\end{aligned}
$$

Hence

$$C + P^A = P^{A_1} \cdot D_1 + \cdots + P^{A_{r-1}} \cdot D_{r-1} + P^A$$

and it follows by induction on r that

$$C + P^A = P^A$$

and hence that $P^A \in \mathscr{H}(+)$.

10.1.7 COROLLARY. For any co-ordinal $A > 0$, W^A is a principal number for addition.

10.1.8 LEMMA. If $A \in \mathscr{H}(+)$ then $A = W^n$ for some n, or for all n, $W^n < A$.

PROOF. If $A \in \mathscr{Q}(+)$ then $A > 1$ and hence by lemma 6.3.1,

$$1 \cdot W = W \leq A.$$

If $A \neq W$, then $W < A$. Now suppose

$$W^n < A\,,$$

then by lemma 6.3.1,

$$W^n \cdot W = W^{n+1} \leqslant A\,.$$

The lemma now follows by induction.

In the same way one proves the following corollary; the details are left to the reader.

10.1.9 COROLLARY. If A, $B \in \mathscr{Q}(+)$ and $B < A$, then either $B \cdot W^n = A$ or for all n, $B \cdot W^n < A$.

10.1.10 THEOREM. The collection $\mathscr{H}(+)$ of principal numbers for addition is strictly ω^ω-unique.

PROOF. By corollary 10.1.7, all the co-ordinals of the form W^n for $n > 0$ are principal numbers for addition. By lemma 10.1.8, these are the only principal numbers for addition with classical ordinals $< \omega^\omega$. Hence $\mathscr{H}(+)$ is ω^ω-unique. By corollary 10.1.7, W^W and W^V are also principal numbers for addition of ordinal ω^ω. But by theorem 8.5.3,

$$W^W = W^V \Rightarrow W = V\,,$$

which is impossible; so $\mathscr{H}(+)$ is strictly ω^ω-unique.

Corollary 10.1.7 above may be regarded as the recursive analogue of the classical result that ω^α is a (classical) principal number for addition for any $\alpha > 0$. However, not every principal number for addition is of the form W^A as we shall show later in chapter 12.

10.2 In this section we establish the ω^{ω^ω}-uniqueness of the collection of principal numbers for multiplication.

10.2.1 LEMMA. Every principal number for multiplication is a principal number for addition, but not conversely.

PROOF. Suppose $P \in \mathscr{H}(\cdot)$, then $2 < P$ and hence $2P = P$. Now by lemma 7.4.1 it follows that 2^W divides P, but by theorem 7.4.2, $2^W = W$, whence W divides P and $W \leq P$. By lemma 6.3.1 we therefore have

$$1 + P = P\,.$$

Suppose, then, that $0<A<P$, therefore

$$\begin{aligned} P = AP &= A(1+P) \\ &= A + AP \\ &= A + P \end{aligned}$$

and we conclude $P \in \mathscr{H}(+)$.

The converse is false since W^2 is a principal number for addition by corollary 10.1.7 but ω^2 is not a (classical) principal number for multiplication and so $W^2 \notin \mathscr{H}(\cdot)$.

10.2.2 COROLLARY. If $P \in \mathscr{H}(\cdot)$, then W divides P.

10.2.3 THEOREM. If $P_1, P_2 \in \mathscr{H}(\cdot)$, then

$$P_1 + P_2 \in \mathscr{H}(\cdot) \Leftrightarrow P_1 + P_2 = P_2 \Leftrightarrow P_1 < P_2 .$$

PROOF. Immediate from lemma 10.2.1 and theorem 10.1.4.

10.2.4 THEOREM. If $P_1, P_2 \in \mathscr{H}(\cdot)$, then

$$P_1 P_2 \in \mathscr{H}(\cdot) \Leftrightarrow P_1 P_2 = P_2 \Leftrightarrow P_1 < P_2 .$$

PROOF. The implications from right to left are obvious. Now suppose $P_1 \cdot P_2 \in \mathscr{H}(\cdot)$, then $P_1 < P_1 P_2$ and hence

$$P_1 (P_1 P_2) = P_1 P_2 .$$

By theorem 8.3.5 it now follows that

$$P_1 P_2 = P_2 .$$

10.2.5 LEMMA. If $B+A=A$, then $(A+B) \cdot W = A \cdot W$.

PROOF.

$$\begin{aligned} (A+B).W &= A + (B+A).W && \text{by theorem 6.2.3.(iii),} \\ &= A + A.W && \text{by hypothesis,} \\ &= A.W && \text{by theorem 5.2.6.(i)} \end{aligned}$$

10.2.6 LEMMA. If $P \in \mathscr{H}(+)$, $D<P$ and $B<A$, then

$$P^B \cdot D + P^A = P^A .$$

PROOF. By corollary 5.2.3, $B<A \Rightarrow B+1 \leq A$, hence

$$\begin{aligned} P^B \cdot D + P^A &= P^B(D + P^{A-B}) && \text{where } A - B \geq 1, \\ &= P^B(D + P + E) && \text{for some } E, \\ &= P^B(P + E) && \text{since } P \in \mathscr{H}(+), \\ &= P^B P^{A-B} \\ &= P^A. \end{aligned}$$

10.2.7 THEOREM. If $P \in \mathscr{H}(\cdot)$, then $P^A \in \mathscr{H}(\cdot)$ if, and only if, $A=1$ or $A \in \mathscr{H}(+)$.

PROOF. If $P^A \in \mathscr{H}(\cdot)$ and $A \neq 1$, then by theorem 8.5.5.(ii),

$$B < A \Rightarrow P^B < P^A$$

and hence

$$P^A = P^B P^A = P^{B+A}.$$

Then by theorem 8.5.3. we have

$$A = B + A,$$

whence $A \in \mathscr{H}(+)$.

Conversely, suppose $A \in \mathscr{H}(+)$ and $0<C<P^A$. By theorem 9.2.7,

$$C = P^{A_1} \cdot D_1 + \cdots + P^{A_r} \cdot D_r = P^{A_1} \cdot D_1 + B, \quad \text{say},$$

where $A>A_1>\cdots>A_r$ and $0<D_i<P$ for $1 \leq i \leq r$.

By theorem 10.2.1, $P \in \mathscr{H}(+)$ and hence by lemma 10.2.6 (used $r-1$ times)

$$C + P^{A_1} \cdot D_1 = P^{A_1} \cdot D \cdot 2$$

and therefore, by corollary 3.2.9,

$$B + P^{A_1} \cdot D_1 = P^{A_1} \cdot D_1. \tag{1}$$

Now, by corollary 10.2.2, W divides P, say

$$P = WE,$$

therefore

$$\begin{aligned} C \cdot P^A &= C \cdot W \cdot E \cdot P^{A-1} \\ &= (P^{A_1} \cdot D_1 + B) \cdot W \cdot E \cdot P^{A-1} \\ &= P^{A_1} \cdot D_1 \cdot W \cdot E \cdot P^{A-1} && \text{by lemma 10.2.5 and (1)}, \\ &= P^{A_1} \cdot D_1 \cdot P \cdot P^{A-1} \end{aligned}$$

$$= P^{A_1+A} \qquad \text{since } P \in \mathscr{H}(\cdot),$$

$$= P^{A} \qquad \text{since } A \in \mathscr{H}(+).$$

This completes the proof.

10.2.8 COROLLARY. For any co-ordinal A, W^{W^A} is a principal number for multiplication.

PROOF. Immediate from theorem 10.2.7 and corollary 10.1.7, for $A>0$. That W is a principal number for multiplication follows at once from theorem 6.2.2.(vi).

10.2.9 LEMMA. If $A \in \mathscr{Q}(\cdot)$, then $A = W^{W^n}$ for some n or, for all n, W^{W^n} divides A and $W^{W^n} < A$.

PROOF. If $A \in \mathscr{Q}(\cdot)$, then $2 < A$ and hence $2A = A$, whence by lemma 7.4.1,

$$2^W = W \text{ divides } A.$$

Now suppose

$$W^{W^n} < A,$$

then

$$(W^{W^n})\,A = A\,, \quad (W^{W^n})^W = W^{W^{n+1}} \text{ divides } A \quad \text{and} \quad W^{W^{n+1}} \leq A\,.$$

The lemma now follows by induction.

In exactly similar fashion one proves the following corollary; we leave the details to the reader.

10.2.10 COROLLARY. If A, $B \in \mathscr{Q}(\cdot)$ and $B < A$, then $B^{W^n} = A$ for some n or, for all n, B^{W^n} divides A and $B^{W^n} < A$.

10.2.11 THEOREM. The collection, $\mathscr{H}(\cdot)$, of principal numbers for multiplication is strictly ω^{ω^ω}-unique.

PROOF. By corollary 10.2.8, all the co-ordinals of the form W^{W^n} are principal numbers for multiplication. By lemma 10.2.9, these are the only principal numbers for multiplication with (classical) ordinals $< \omega^{\omega^\omega}$. Hence $\mathscr{H}(\cdot)$ is ω^{ω^ω}-unique. By corollary 10.2.8, W^{W^W} and W^{W^V} are also principal numbers for multiplication of ordinal ω^{ω^ω}. But by theorem 8.5.3,

$$W^{W^W} = W^{W^V} \Rightarrow W^W = W^V \Rightarrow W = V\,,$$

which is impossible. Hence $\mathscr{H}(\cdot)$ is strictly ω^{ω^ω}-unique.

Corollary 10.2.8 above may be regarded as the recursive analogue of the result that ω^{ω^α} is a (classical) principal number for multiplication. However, not every principal number for multiplication is of the form W^{W^A}. This follows at once from the fact, remarked above and proved in chapter 12, that there is a $P \in \mathscr{H}(+)$ such that $P \neq W^A$ and theorem 10.2.7.[102]

10.3 Since if $P \in \mathscr{H}(+)$, $\mathscr{P}(P)$ is closed under addition, and similarly for $P \in \mathscr{H}(\cdot)$, $\mathscr{P}(P)$ is closed under multiplication and also (by theorem 10.2.1) addition we have the following theorem.

10.3.1 Theorem. If $\begin{matrix} P \in \mathscr{H}(+) \\ P \in \mathscr{H}(\cdot) \end{matrix}$ then

$$\begin{matrix} \langle \mathscr{P}(P), + \rangle \\ \langle \mathscr{P}(P), +, \cdot \rangle \end{matrix} \quad \text{is isomorphic to} \quad \begin{matrix} \langle |P|, \oplus \rangle \\ \langle |P|, \oplus, \odot \rangle \end{matrix} \quad \text{by the map} \quad |\ |$$

(where $\oplus$ and $\odot$ are (classical) addition and multiplication of ordinals).

10.3.2 Corollary, If $P \in \mathscr{H}(+)$ $[P \in \mathscr{H}(\cdot)]$ then $|P| = \omega^\alpha$ $[|P| = \omega^{\omega^\alpha}]$ for some ordinal α.

These results, of course, also apply to any P satisfying the weaker condition (3) of § 8.1 (and its multiplicative analogue in the case of multiplication). The useful fact about taking $P \in \mathscr{H}(+)$ [or $\mathscr{H}(\cdot)$] is that the path can be extended to a longer closed path.

10.3.3 Theorem. For every denumerable ordinal α there is a principal number for addition (for multiplication) P such that $|P| \geq \alpha$.

Proof. Immediate from corollaries 10.1.7 and 10.2.8.

Later (chapter 12) we shall strengthen these results and prove that there exist paths closed under addition and multiplication (and exponentiation) which are of length $\aleph_1$ i.e. "as long as possible". As an immediate application of theorem 10.3.1 we have the following definition and theorem.

10.3.4 Definition. A co-ordinal A is said to be *indecomposable* (with respect to multiplication) if

$$A = BC \quad \text{implies} \quad B = A \quad \text{or} \quad C = A.$$

10.3.5 THEOREM (UNIQUE FACTORIZATION THEOREM). If $1 < B < W^{W^A}$ for some co-ordinal A, then B is uniquely expressible as a finite product of indecomposable factors such that if E, F are two successive factors of B with E preceding F, then

(i) if $E = C+1$, then F is of the form $F = D+1$,

(ii) if E, F are finite, then $E \geq F$,

(iii) if E, F are both not of the form $C+1$, then $E \geq F$,

(iv) if $E = C+1$ for some infinite C, then $C = W^D$ where $0 < D < W^A$,

(v) if $E \neq C+1$ for any C but E is infinite, then $E = W^{W^D}$ where $D < A$.

PROOF. Immediate from theorem 10.3.1 and the classical unique factorization theorem (see e.g. BACHMANN, 1955, Satz 2, p. 88).

10.3.6 COROLLARY. The infinite indecomposable co-ordinals $< W^{W^A}$ are precisely those of the forms

$$W^{W^B} \quad \text{and} \quad W^C + 1 \quad \text{where} \quad B < A, 0 < C < W^A.$$

10.3.7 COROLLARY. The only principal numbers for multiplication which are $< W^{W^A}$ are those of the form W^{W^B} for some $B < A$.

CHAPTER 11

E-NUMBERS[111]

11.1 The analogues of the classical arithmetic laws for addition and multiplication do not, as we have seen, carry over to co-ordinals in general. In the case of exponentiation, however, the situation is much more agreeable and in this chapter we determine explicitly all the principal numbers for exponentiation and show that analogues of differing classical methods of defining ε-numbers yield the same collection of principal numbers for exponentiation. As a corollary we get the result that the collection of principal numbers for exponentiation is strictly ε_ω-unique.[112]

11.1.1 THEOREM. $\mathscr{H}(\exp) \subset \mathscr{H}(\cdot)$.

PROOF. Suppose $P \in \mathscr{H}(\exp)$, then by lemma 8.5.8,

$$\text{if} \quad A < P, \quad \text{then} \quad A^A < P$$

and hence, for $1 < A < P$,

$$A^{AP} = (A^A)^P = P = A^P.$$

Therefore, for such A,

$$AP = P \quad \text{by theorem 8.5.3}$$

and hence $P \in \mathscr{H}(\cdot)$.

Now $W^W \in \mathscr{H}(\cdot)$ by corollary 10.2.8 and $W < W^W < W^{(W^W)}$ hence $W^W \notin \mathscr{H}(\exp)$ so $\mathscr{H}(\exp) \neq \mathscr{H}(\cdot)$. This completes the proof.

We now describe a construction told us by Parikh which assigns, in a natural way, a well-ordering of type ε_α to a given well-ordering of type α.

11.1.2 Definition. If A is a well-ordering, then we set

$$s(\mathsf{A}) = \{\langle x+2, y+2\rangle : \langle x, y\rangle \in \mathsf{A}\} \quad \text{and}$$
$$t(\mathsf{A}) = \{\langle 0,0\rangle, \langle 0,1\rangle, \langle 1,1\rangle\} \mathbin{\hat{+}} s(\mathsf{A}).$$

Next we define E(A) to be the smallest set $\boldsymbol{E}$ of ordered pairs of natural numbers satisfying 1)–3) below

1) If $\langle x, y\rangle \in t(\mathsf{A})$, then

$$\left\langle e\begin{pmatrix}x\\1\end{pmatrix}, e\begin{pmatrix}y\\1\end{pmatrix}\right\rangle \in \boldsymbol{E},$$

2) If[113] $x_1 \geq_{\boldsymbol{E}} \cdots \geq_{\boldsymbol{E}} x_r$, where $r \geq 1, m_1, \ldots, m_r \geq 1$ and either $r > 1$ or $m_1 > 1$ or x_1 is not of the form $e\begin{pmatrix}x+2\\1\end{pmatrix}$ and analogous conditions hold for $y_1, \ldots, y_s, s, n_1, \ldots, n_s$, then

$$\left\langle e\begin{pmatrix}x_1 \ldots x_r 1\\ m_1 \ldots m_r 0\end{pmatrix}, e\begin{pmatrix}y_1 \ldots y_s 1\\ n_1 \ldots n_s 0\end{pmatrix}\right\rangle \in \boldsymbol{E} \quad \text{if, and only if,}$$

$$(\exists t)\,\{t < s \;\&\; (\forall q)(q \leq t \Leftrightarrow (0 \leq i \leq q \Rightarrow j(x_i, m_i) = j(y_i, n_i)))$$
$$\&\; [t = r \,.\vee.\, \{t < r \;\&\; [((\langle x_{t+1}, y_{t+1}\rangle \in \boldsymbol{E} \;\&\; x_{t+1} \neq y_{t+1})$$
$$\vee\, (x_{t+1} = y_{t+1} \;\&\; m_{t+1} \leq n_{t+1})]\}]\},$$

3) If $x = e\begin{pmatrix}x_1 \ldots x_r 1\\ n_1 \ldots n_r 0\end{pmatrix}$ and $\langle x, x\rangle \in \boldsymbol{E}$ and $y = e\begin{pmatrix}y_0\\1\end{pmatrix}$

and $\langle y, y\rangle \in \boldsymbol{E}$, then

$$\langle x, y\rangle \in \boldsymbol{E} \quad \text{if} \quad \langle x_i, y\rangle \in \boldsymbol{E} \quad \text{for} \quad 1 \leq i \leq r,$$
$$\langle y, x\rangle \in \boldsymbol{E} \quad \text{otherwise}.$$

Clearly E(A) and C‘E(A) are uniformly primitive recursive in A.

11.1.3 Theorem. If A is a well-ordering of type α, then E(A) is a well-ordering of type ε_α.

Proof. Let φ be the map from $\varepsilon_\alpha = \{\beta : \beta < \varepsilon_\alpha\}$ into C‘E(A) given by

$$\varphi(0) = e\begin{pmatrix}0\\1\end{pmatrix},$$

$$\varphi(\omega) = e\begin{pmatrix}1\\1\end{pmatrix},$$

$$\varphi(\varepsilon_\gamma) = e\binom{x+2}{1} \qquad \text{where } \gamma < \alpha,\ |x|_{\mathsf{A}} = \gamma,$$

$$\varphi(\beta) = e\begin{pmatrix} x_1 \dots x_r 1 \\ m_1 \dots m_r 0 \end{pmatrix} \quad \text{where } \beta = \omega^{\beta_1}\cdot m_1 + \dots + \omega^{\beta_r}\cdot m_r,$$

β is not an ε-number, ω or 0, $\varphi(\beta_i) = x_i$ and

$\beta_1 > \dots > \beta_r$.

We prove by transfinite induction that φ is well-defined and maps ε_α one-one onto C'E(A) in an order preserving way; from this the theorem follows at once.

φ is well-defined since if $\beta < \varepsilon_\alpha$ then either (i) exactly one of $\beta = 0$, $\beta = \omega$, $\beta = \varepsilon_\gamma$ for some unique $\gamma < \alpha$ holds or (ii) $\beta = \omega^{\beta_1}\cdot m_1 + \dots + \omega^{\beta_r}\cdot m_r$ and in case (ii) the β_i, m_i are uniquely determined if we specify $\beta_1 > \dots > \beta_r$. (Note that all the β_i are $< \beta$ in this case.) That φ is one-one is clear from the uniqueness of the Cantor normal form. Since every ordinal $< \varepsilon_\alpha$ is expressible in the forms given above it follows that φ is onto and the order preserving property of φ is obvious from definition 11.1.2 and the observation that

$$\omega^{\beta_1}\cdot m_1 + \dots + \omega^{\beta_r}\cdot m_r < \varepsilon_\gamma \Leftrightarrow \beta_i < \varepsilon_\gamma \quad \text{for} \quad 1 \le i \le r.$$

This completes the proof.

A straightforward calculation shows that $\mathsf{A} \simeq \mathsf{B}$ if, and only if, $\mathrm{E}(\mathsf{A}) \simeq \mathrm{E}(\mathsf{B})$ but we prove a stronger result.

Notation. We write $\mathsf{A} \le_p \mathsf{B}$ if

(i) $\mathsf{A} \le \mathsf{B}$,

(ii) p is partial recursive, $\delta p \supseteq$ C'A, $\rho p \subseteq \{0, 1\}$,

(iii) if $x \in$ C'B and $p(x)$ is defined, then

$$p(x) = 0 \Leftrightarrow x \in \text{C'A} \;.\&.\; p(x) = 1 \Leftrightarrow x \in \text{C'B} - \text{C'A}.$$

11.1.4 LEMMA. If $A \le B$ then for each $\mathsf{B} \in B$ there exists $\mathsf{A} \in A$ and p such that (i) $\delta p \supseteq$ C'B and (ii) $\mathsf{A} \le_p \mathsf{B}$.

PROOF. Let $\mathsf{B} \in B$, then since $A \le B$ there exists $\mathsf{A} \in A$ such that $\mathsf{A} \le \mathsf{B}$. Without loss of generality we may assume $\mathsf{B} = [\mathrm{B}]$ and therefore, since $\mathsf{A} \le \mathsf{B}$, that $\mathsf{A} = [\mathrm{A}]$ for some $\mathrm{A} \subseteq \mathrm{B}$. By theorems 2.1.2, 3 there exists p such that $\delta p = \mathrm{Seq}$ and $\mathsf{A} \le_p \mathsf{B}$.

11.1.5 LEMMA. $\mathsf{A} \leq \mathsf{B} \Leftrightarrow \mathrm{E}(\mathsf{A}) \leq \mathrm{E}(\mathsf{B})$.

PROOF. By lemma 11.1.4, if $\mathsf{A} \leq \mathsf{B}$ then there exists p such that

$$\delta p \supseteq \mathrm{C}^{\backprime}\mathsf{B} \quad \text{and} \quad \mathsf{A} \leq_p \mathsf{B}.$$

Let q be the partial recursive function defined by

$$q\left(e\begin{pmatrix}0\\1\end{pmatrix}\right) = q\left(e\begin{pmatrix}1\\1\end{pmatrix}\right) = 0,$$

$$q\left(e\begin{pmatrix}x+2\\1\end{pmatrix}\right) = p(x),$$

$$q\left(e\begin{pmatrix}x_1 \dots x_r 1\\ m_1 \dots m_r 0\end{pmatrix}\right) = q(x_1) \text{ if } r > 1 \text{ or } m_1 > 1 \text{ or } x_1 \neq e\begin{pmatrix}y+2\\1\end{pmatrix} \text{ for all } y$$

and undefined otherwise.

(q is an extension of $p\varphi_{\mathsf{B}}\varphi_{\mathsf{A}}^{-1}$ where φ_{A}, φ_{B} are the maps φ from $\varepsilon_{|\mathsf{A}|}$, $\varepsilon_{|\mathsf{B}|}$ onto $\mathrm{C}^{\backprime}\mathrm{E}(\mathsf{A})$, $\mathrm{C}^{\backprime}\mathrm{E}(\mathsf{B})$, respectively, defined as in the proof of theorem 11.1.3.)

By transfinite induction (as in the proof of theorem 11.1.3.) one readily verifies that

$$\delta q \supseteq \mathrm{C}^{\backprime}\mathrm{E}(\mathsf{B}) \quad \text{and} \quad \mathrm{E}(\mathsf{A}) \leq_q \mathrm{E}(\mathsf{B});$$

we leave the details to the reader.

Conversely, suppose $\mathrm{E}(\mathsf{A}) \leq \mathrm{E}(\mathsf{B})$, then by the lemma for some q, $\mathrm{E}(\mathsf{A}) \leq_q \mathrm{E}(\mathsf{B})$ and $\delta q \supseteq \mathrm{C}^{\backprime}\mathrm{E}(\mathsf{B})$. Now from the definition of $\mathrm{E}(\mathsf{A})$ we have

$$\mathsf{A} = \left\{\langle x, y\rangle : \left\langle e\begin{pmatrix}x+2\\1\end{pmatrix}, e\begin{pmatrix}y+2\\1\end{pmatrix}\right\rangle \in \mathrm{E}(\mathsf{A})\right\}$$

and similarly B may be obtained from $\mathrm{E}(\mathsf{B})$. Clearly

$$\mathsf{A} \leq_p \mathsf{B} \quad \text{if we put} \quad p(x) = q\left(e\begin{pmatrix}x+2\\1\end{pmatrix}\right).$$

11.1.6 THEOREM. $\mathsf{A} \simeq \mathsf{B} \Leftrightarrow \mathrm{E}(\mathsf{A}) \simeq \mathrm{E}(\mathsf{B})$,

PROOF. Similar to the above: left to the reader.

This theorem justifies the following definition.

11.1.7 Definition. $\mathrm{E}(A) = \mathrm{COT}(\mathrm{E}(\mathsf{A}))$ for any $\mathsf{A} \in A$. A co-ordinal of the form $\mathrm{E}(A)$ is said to be an E-*number*.

11.1.8 Corollary. $A \leq B \Leftrightarrow \mathrm{E}(A) \leq \mathrm{E}(B)$.

Proof. Immediate from lemma 11.1.5.

11.2 The following is the main theorem of this chapter.

11.2.1 Theorem. The following statements are equivalent:

(i) X is an infinite principal number for exponentiation,

(ii) $2^X = X$,

(iii) $X = W$ or $W^X = X$,

(iv) $X = W$ or X is an E-number.

Proof. (i)⇒(ii) follows from the definition of principal numbers. For (ii)⇒(iii) we need a lemma.

11.2.2 Lemma. $2^X = X \Rightarrow X = WD$ for some $D > 0$.

We first sketch the idea of the proof. Since $p : 2^{\mathsf{X}} \simeq \mathsf{X}$ for some p, we can represent each $|x|$ for $x \in \mathrm{C}^{\text{‘}}\mathsf{X}$ in the form

$$|x| = 2^{|x_1|} + \cdots + 2^{|x_s|} + 2^{m_1} + \cdots + 2^{m_r},$$

where $m_1 > \cdots > m_r$ are the (only) finite exponents. Given such an expression then $|x| + 1$ is obtained by adding 2^0 and reducing the result so that we again get a unique expression. Clearly this addition can only affect the finite exponents and so the problem is to determine explicitly and effectively how "far in" the 2^0 has an effect on the exponents. The calculation of $t = t(x)$, below, does just this. The function ψ enumerates the elements of $\mathrm{C}^{\text{‘}}\mathsf{X}$ corresponding to the finite ordinals. Using these two functions it easily follows that $\mathsf{X} \simeq \mathsf{W} \cdot \mathsf{D}$ where D is obtained from the numbers which do not represent successor numbers in X.

Proof. Let $\mathsf{X} \in X$, then by hypothesis there exists p such that

$$p : 2^{\mathsf{X}} \simeq \mathsf{X}.$$

Let ψ be the partial recursive function (uniformly recursive in p) defined by

$$\psi(0) = p(e(\mathbf{0})),$$

$$\psi(2^{m_1} + \cdots + 2^{m_r}) = p\left(e\begin{pmatrix}\psi(m_1) \ldots \psi(m_r)\\ 1 \quad \ldots \quad 1\end{pmatrix}\right),$$

where $m_1 > \cdots > m_r$.
Since every integer >0 has a unique dyadic expansion, ψ is well-defined.
If $x = e\begin{pmatrix}x_0 \ldots x_r\\ 1 \ \ldots \ 1\end{pmatrix}$ let

$$t(x) = \mu_z\{0 \le z \le r+1 \ \& \ (\forall i)\,(z \le i \le r \Rightarrow x_i = \psi(r-i))\}$$

and be undefined for all other x and, further, writing t for $t(x)$ since no confusion will arise, let

$$\begin{aligned} s(x) &= e\begin{pmatrix}\psi(r+1)\\ 1\end{pmatrix} && \text{if} \quad t = 0,\\ &= e\begin{pmatrix}x_0 \ldots x_{t-1} & \psi(r+1-t)\\ 1 \ldots 1 & 1\end{pmatrix} && \text{if} \quad r \ge t > 0,\\ &= e\begin{pmatrix}x_0 \ldots x_r & \psi(0)\\ 1 \ldots 1 & 1\end{pmatrix} && \text{if} \quad t = r+1,\end{aligned}$$

be undefined for x not of the above form.

Finally, let

$$\mathsf{D} = \{\langle x, y\rangle : \langle x, y\rangle \in 2^{\mathsf{X}} \ \& \ (\forall z)\,(x \neq s(z) \ \& \ y \neq s(z))\},$$

then the reader will readily verify that

$$q : \mathsf{W}\cdot\mathsf{D} \simeq \mathsf{X},$$

where $q(j(n, d)) = p(s^n(d))$.

11.2.3 Lemma. $2^X = X \Rightarrow W^X = X$ or $X = W$.

Proof. If $2^X = X$, then $X = WD$ for some D by the previous lemma. Hence $2^{WD} = WD = X$. But $2^{WD} = (2^W)^D = W^D$ by theorem 7.4.2 and hence $WD = W^D$. Now $D > 0$, since $X > 0$, hence for some E, $D = 1 + E$ and

$$W^D = W^{1+E} = W\cdot W^E = WD.$$

By theorem 8.3.5 it follows that

$$W^E = D = 1 + E.$$

If $E = 0$; then $X = W$; otherwise $E > 0$ and by corollary 10.1.7 we have

$$1 + E = W^E = 1 + W^E.$$

Therefore, $E = W^E = D$ and

$$D = W^D = X = W^X.$$

Thus (ii)⇒(iii). Using a related sort of argument to that used to prove lemma 11.2.2 we prove (iii)⇒(iv). The essential point is that any number not representing an ε-number (in the ordering X) is expressible in Cantor normal form in terms of numbers representing (strictly) smaller ordinals.

11.2.4 LEMMA. If $W^X = X$, then $X = \mathrm{E}(A)$ for some A.

PROOF. Suppose $W^X = X$ and $\mathsf{X} \in X$, then there exists a recursive isomorphism p such that

$$p : \mathsf{X} \simeq \mathsf{W}^{\mathsf{X}}.$$

We may assume $\min(\mathsf{X}) = 0$.

Let

$$\mathsf{A} = \left\{ \langle x, y \rangle \in \mathsf{X} : e\binom{x}{1} = p(x) \;\&\; e\binom{y}{1} = p(y) \right\}$$

and let q be the partial recursive function defined by

$$q\left(e\binom{0}{1}\right) = e(0) \qquad (1)$$

$$q\left(e\binom{1}{1}\right) = e\begin{pmatrix} e\binom{0}{1} \\ 1 \end{pmatrix} \qquad (2)$$

$$q\left(e\binom{x+2}{1}\right) = e\binom{x}{1} \quad \text{if} \quad e\binom{x}{1} = p(x) \qquad (3)$$

$$q\left(e\begin{pmatrix} x_1 \dots x_r 1 \\ m_1 \dots m_r 0 \end{pmatrix}\right) = e\begin{pmatrix} p^{-1}q(x_1) \dots p^{-1}q(x_r) \\ m_1 \quad \dots \quad m_r \end{pmatrix} \qquad (4)$$

and undefined otherwise. (5)

Then it is easily proved that $\delta q \supseteq \mathrm{C}\text{'}\mathrm{E}(\mathsf{A})$ (by transfinite induction) and the q maps $\mathrm{C}\text{'}\mathrm{E}(\mathsf{A})$ into $\mathrm{C}\text{'}\mathsf{W}^{\mathsf{X}}$. q is one-one. We prove this by induction on the maximum number n of applications of the cases (1)–(4) above in the computation of the $q(x_i)$ necessary to compute $q(x)$. If $n = 0$, then only (1)–(3) arise. The only difficulty is if cases (2), (3) conflict. But $e\binom{0}{1}$

represents the ordinal 1 in W^{X} so the right hand side of (2) represents ω.

On the other hand $e\begin{pmatrix} x \\ 1 \end{pmatrix}$, if it is in C'E(A), represents a fixed point of the isomorphism induced between the ordinals by p by the condition in (3) and therefore represents an ε-number in W^{X}, so there is in fact no conflict of requirements. Now suppose

$$q(x) = q(y) = e\begin{pmatrix} a_1 \dots a_r \\ m_1 \dots m_r \end{pmatrix},$$

then the a_i, m_i and r are uniquely determined and each $a_i = q(x_i)$ for some x_i by condition (5). Moreover, the x_i are uniquely determined by the induction hypothesis. Thus we see that r, the x_i and the m_i are all uniquely determined and since e is one-one we conclude $x = y$.

Finally, we leave the reader to prove by transfinite induction that

$$q : \mathrm{E}(\mathsf{A}) \simeq \mathsf{W}^{\mathsf{X}},$$

whence $\mathrm{E}(\mathsf{A}) \simeq \mathsf{X}$. Putting $A = \mathrm{COT}(\mathsf{A})$ yields the required result.

Now we show (iv)⇒(iii).

11.2.5 Lemma. $W^{\mathrm{E}(A)} = \mathrm{E}(A)$.

Proof. Suppose $\mathsf{A} \in A$, then let f be the partial recursive function defined by

$$f\left(e\begin{pmatrix} 0 \\ 1 \end{pmatrix}\right) = e(0),$$

$$f\left(e\begin{pmatrix} 1 \\ 1 \end{pmatrix}\right) = e\begin{pmatrix} e\begin{pmatrix} 0 \\ 1 \end{pmatrix} \\ 1 \end{pmatrix},$$

$$f\left(e\begin{pmatrix} x+2 \\ 1 \end{pmatrix}\right) = e\begin{pmatrix} e\begin{pmatrix} x+2 \\ 1 \end{pmatrix} \\ 1 \end{pmatrix},$$

$$f\left(e\begin{pmatrix} x_1 \dots x_r 1 \\ m_1 \dots m_r 0 \end{pmatrix}\right) = e\begin{pmatrix} x_1 \dots x_r \\ m_1 \dots m_r \end{pmatrix},$$

and undefined otherwise.

Then it is readily verified as in the preceding proof that

$$f : \mathrm{E}(\mathsf{A}) \simeq \mathsf{W}^{\mathrm{E}(\mathsf{A})},$$

from which the lemma follows.

Finally we show that (iii)$\Rightarrow$(ii)&(i). This requires a number of lemmata.

11.2.6 LEMMA. $BA = A \Rightarrow (AB)^W = A^W$.

PROOF.
$$\begin{aligned} (AB)^W &= (AB)^{2+W} \\ &= (AB)(AB)(AB)^W \\ &= A(BA)\,B(AB)^W \\ &= A(AB)(AB)^W \quad \text{by the hypothesis,} \\ &= A(AB)^W. \end{aligned}$$
By lemma 7.4.1, it follows that A^W divides $(AB)^W$ and, since $|A^W| = |(AB)^W|$ (by e.g. BACHMANN, 1955, p. 57) it follows from theorem 8.3.3.(iii) that $A^W = (AB)^W$.

11.2.7 LEMMA. $B + A = A \,\&\, A = WD + n \Rightarrow (A+B)^W = A^W$

PROOF. We leave the trivial case $D = 0$ to the reader. We first observe that if $B + A = WD$ then

$$\begin{aligned} (A+B)\,A^W &= (A+B)(WD)^W \\ &= ((A+B)\,W)\,D(WD)^W \\ &= (AW)\,D(WD)^W \qquad \text{by theorem 8.2.2 with } AW \text{ for } C, \\ &= (WDW)\,D(WD)^W \\ &= (WD)^{2+W} = (WD)^W = A^W. \end{aligned}$$

By lemma 7.4.1 it follows that $(A+B)^W$ divides A^W and hence (by e.g. BACHMANN, 1955, p. 57 and theorem 8.3.3.(iii)) that

$$A^W = (A+B)^W.$$

Now suppose $A = WD + n$ where $n > 0$, then

$$(A+B)^W = (WD + n + B)^W.$$

Hence

$$(A+B)^W = (WD + B')^W$$

where

$$B' = n + B.$$

If $B+A=A$, then

$$B + WD + n = WD + n$$

and hence by theorem 5.1.6,

$$B + WD = WD\,.$$

Therefore

$$B' + WD = n + B + WD = n + WD = WD$$

by lemma 7.4.1 (since we are assuming $D \neq 0$). Putting $A' = WD$ we immediately obtain from the first part of the proof

$$(A' + B')^W = A'^W\,,$$

i.e.

$$(WD + n + B)^W = (WD)^W\,,$$

i.e.

$$(A + B)^W = (WD)^W\,. \tag{6}$$

Finally $n + WD = WD$ and hence putting $B=n$, $A=WD$ in the first part of the proof we get

$$(WD + n)^W = (WD)^W\,. \tag{7}$$

Combining (6) and (7) gives the desired result.

11.2.8 LEMMA. $W^X = X \Rightarrow X$ is a principal number for addition and multiplication.

PROOF. This follows at once from corollaries 10.1.7 and 10.2.8.

11.2.9 LEMMA. $W^X = X \,\&\, 1 < Y < X \Rightarrow Y^X = X$.

PROOF. By theorem 9.2.8 the hypothesis implies

$$Y = W^{X_1} \cdot n_1 + \cdots + W^{X_r} \cdot n_r$$

where $r \geq 1$, $X > X_1 > \cdots > X_r$ and the n_i are finite, non-zero.

If $r=1$, $X_1=0$, then $Y = n_1 > 1$, hence

$$Y^X = n_1^X = n_1^{W^X} = n_1^{W \cdot W^X} = W^{W^X} = X\,,$$

since W^X is a principal number for multiplication and W is a principal number for exponentiation by theorem 7.4.2. In any other case we have

$$\begin{aligned} Y^X = Y^{W^X} &= (Y^W)^{W^X} = ((W^{X_1} \cdot n_1)^W)^{W^X} && \text{by lemma 11.2.7,} \\ &= ((W^{X_1})^W)^{W^X} && \text{by lemma 11.2.6,} \\ &= W^{X_1 W W^X} = W^{X_1 W^X} = W^{W^X} && \text{by lemma 11.2.8,} \\ &= X\,. \end{aligned}$$

Putting $Y=2$ we get (iii)⇒(ii) and the stated version of the last lemma gives (iii)⇒(i), so the proof of the theorem is complete.

11.2.10 Corollary. The collection $\mathscr{H}(\exp)$ of all principal numbers for exponentiation is strictly ε_ω-unique.

Proof. By theorem 11.2.1 every principal number $\neq W$ is of the form $\mathrm{E}(A)$ for some co-ordinal A. By theorem 11.1.3 if $|\mathrm{E}(A)|<\varepsilon_\omega$ then $|\mathrm{E}(A)|=\varepsilon_n$ for some n and consequently $|A|=n$ which implies $A=n$. So $\mathscr{H}(\exp)$ is ε_ω-unique. Now consider $\mathrm{E}(W)$ and $\mathrm{E}(V)$; by corollary 11.1.8, $\mathrm{E}(W)=\mathrm{E}(V)$ implies $W=V$ which is a contradiction, thus $\mathscr{H}(\exp)$ is strictly ε_ω-unique.

11.2.11 Theorem. If $P\in\mathscr{H}(\exp)$, then

$$\langle \mathscr{P}(P), +, \cdot, \exp\rangle \text{ is isomorphic to}$$
$$\langle |P|, \oplus, \odot, exp\rangle \text{ by the map } |\ |$$

(where $\oplus$, $\odot$, *exp* denote addition, multiplication and exponentiation of ordinals).

Proof. Since $A, B<\mathrm{P}$ imply $A^B<P$ the theorem follows at once from theorem 10.3.1.

CHAPTER 12

INFINITE SERIES

12.1 In this chapter we introduce constructive sequence types (C.S.T.s) in the spirit of ACZEL (1966). We first of all prove results for C.S.T.s analogous to those for C.O.T.s. and then we use them to prove that there are no non-trivial least upper bounds for collections of co-ordinals but also that there exist paths of co-ordinals of length equal to the first uncountable ordinal which are closed under addition, multiplication and exponentiation.

If C is a linear ordering and to each $i \in \mathsf{C}`\mathsf{C}$ is assigned (not necessarily by any constructive rule) a linear ordering A_i, then we write

$$\boldsymbol{A} = \langle \mathsf{A}_i : \mathsf{C} \rangle \tag{1}$$

for the sequence $\{\mathsf{A}_i : |i|_{\mathsf{C}} < |\mathsf{C}|\}$ and call $\boldsymbol{A}$ a *sequence ordering* (S.O.). If A_i, C are all well-orderings we call $\boldsymbol{A}$ a *sequence well-ordering* (S.W.O.). We shall generally use bold face capitals $\boldsymbol{A}, \boldsymbol{B}, \ldots$ to denote such S.O.s.

We introduce recursive isomorphism of S.O.s. in what we believe is the natural way, but first we make precise what we mean by (classical) isomorphism of S.O.s.

12.1.1 DEFINITION. Let $\boldsymbol{A} = \langle \mathsf{A}_i : \mathsf{C} \rangle$, $\boldsymbol{B} = \langle \mathsf{B}_j : \mathsf{D} \rangle$ be S.O.s, then $\boldsymbol{A}$ is said to be *isomorphic* to $\boldsymbol{B}$ if there exist functions q, $p_i (i \in \mathsf{C}`\mathsf{C})$ such that

(i) $q : \mathsf{C} \sim \mathsf{D}$ and
(ii) for all $i \in \mathsf{C}`\mathsf{C}$, $p_i : \mathsf{A}_i \sim \mathsf{B}_{q(i)}$.

12.1.2 DEFINITION. Let $\boldsymbol{A}, \boldsymbol{B}$ be as in the previous definition; then a pair

$\lambda ixp(i, x)$, $\lambda xq(x)$ is said to be a *recursive isomorphism pair* if
(i) $q:\mathsf{C} \simeq \mathsf{D}$,
(ii) for all $i \in \mathsf{C}`\mathsf{C}$, $\lambda xp(i, x):\mathsf{A}_i \simeq \mathsf{B}_{q(i)}$.
(iii) $\lambda xp(i, x)$ is one-one for all i.
We write $p, q:\boldsymbol{A} \simeq \boldsymbol{B}$ if p, q is such a pair and $\boldsymbol{A} \simeq \boldsymbol{B}$ if such a pair exists and we say that $\boldsymbol{A}$ and $\boldsymbol{B}$ are *recursively isomorphic.*

Note that since we require p to be a partial recursive function of *both* variables the recursive isomorphisms in 12.1.2(ii) are *uniform in i.*

Recursive isomorphism (of S.O.s) is an equivalence relation – this is immediate from the remarks following definition 1.1.4 and the fact that if $\lambda xp(i, x)$ is one-one for all $i \in \mathsf{C}`\mathsf{C}$ then the function p' defined by

$$p'(i, x) = y \Leftrightarrow p(i, y) = x$$

is uniformly recursive in i.

12.1.3 DEFINITION. If $\boldsymbol{A}$ is a S.O. and $\mathscr{A} = \{\boldsymbol{B}:\boldsymbol{B} \simeq \boldsymbol{A}\}$ then $\mathscr{A}$ is said to be a *constructive sequence type* (C.S.T.) and we write $\mathscr{A} = \mathrm{CST}(\boldsymbol{A})$. If $\boldsymbol{A}$ is a sequence well-ordering then $\mathscr{A}$ is said to be a *sequence co-ordinal.*

In dealing with recursive isomorphism pairs we need not have assumed that $\lambda xp(i, x)$ was one-one for all i and not just for the $i \in \mathsf{C}`\mathsf{C}$ (where $\boldsymbol{A}, \boldsymbol{B}, p, q$ are as in definition 12.1.2) for if p were not so then we could define

$$p'(i, x) = p(i, x) \quad \text{if} \quad x = \nu_y(p(i, x) = p(i, y)),$$
$$\text{undefined otherwise,}$$

where ν is Kleene's indefinite description operator (KLEENE, 1952, p. 347). Then, using the fact that $p(i, x)$ is one-one for $i \in \mathsf{C}`\mathsf{C}$,

$$p(i, x) = p'(i, x) \quad \text{for such } i.$$

If $p, q:\boldsymbol{A} \simeq \boldsymbol{B}$ except that condition 12.1.2(iii) is not satisfied, then $p', q:\boldsymbol{A} \simeq \boldsymbol{B}$. But this condition being satisfied for a recursive isomorphism pair p', q means that this pair determines a r.e. sequence ordering $\boldsymbol{A} = \langle \mathsf{A}_j : \mathsf{C} \rangle$ defined by

$$\mathsf{A}_j = (\delta \lambda x p'(j, x))^2 \cap \mathsf{R},$$
$$\mathsf{C} = (\delta q)^2 \cap \mathsf{R}$$

(note that q is one-one on the whole of its domain by definition 1.1.4).

12.2 In chapter 2 we defined

$$\mathsf{A} + \mathsf{B} = \hat{0}\mathsf{A} \mp \hat{1}\mathsf{B}$$

and we now regard this as the sum of the two element sequence $\langle \mathsf{A}, \mathsf{B} \rangle$ in order to generalize the notion of sum to take in the infinite case. Let us define, for $\boldsymbol{A}$ a S.O.

$$\Sigma \boldsymbol{A} = \hat{\mathsf{C}}(\mathsf{A}_i)$$
$$= \{\langle \hat{\imath}(a), \hat{\jmath}(b) \rangle : (i <_{\mathsf{C}} j \;\&\; a \in \mathrm{C}`\mathsf{A}_i \;\&\; b \in \mathrm{C}`\mathsf{A}_j) \quad \text{or}$$
$$(i = j \in \mathrm{C}`\mathsf{C} \;\&\; a \leq_{\mathsf{A}_i} b)\}.$$

12.2.1 LEMMA. If $\boldsymbol{A} = \langle \mathsf{A}_i : \mathsf{C} \rangle$ and $\mathsf{A}_i (i \in \mathrm{C}`\mathsf{C})$ and $\mathsf{C} \subseteq \mathsf{R}$, then $\Sigma \boldsymbol{A} \subseteq \mathsf{R}$.

PROOF. Left to the reader.

12.2.2 THEOREM. There is a partial recursive functional Σ such that if $p, q : \boldsymbol{A} \simeq \boldsymbol{B}$ then

$$\Sigma(p, q) : \Sigma \boldsymbol{A} \simeq \Sigma \boldsymbol{B}, \tag{2}$$

where the $\simeq$ in (2) denotes recursive isomorphism of linear orderings.

PROOF. Set

$$\Sigma(p, q)(x) = \hat{z}(p(y, \bar{y}(x))),$$

where[121]

$$z = q((x)_0) \quad \text{and} \quad y = (x)_0 .$$

Clearly Σ is a partial recursive functional and

$$\Sigma(p, q)(\hat{\imath}(a)) = \widehat{q(i)}(p(i, a)).$$

Suppose $\Sigma(p, q)(x) = t$, then $t = \hat{\imath}(u)$ for some uniquely determined i, u. Since q is one-one, there is a unique z such that $q(z) = i$. Since $\lambda x p(i, x)$ is one-one *for all* i there is a unique v such that $p(z, v) = u$. But $\hat{z}(v) = x$ and hence x is uniquely determined by t so $\Sigma(p, q)$ is one-one everywhere. We leave the reader to check that $\Sigma(p, q)$ satisfies (iii), (iv) of definition 1.1.4.

12.2.3 DEFINITION. If $\mathscr{A}$ is a C.S.T., then $\Sigma \mathscr{A} = \mathrm{COT}(\Sigma \boldsymbol{A})$ for any $\boldsymbol{A} \in \mathscr{A}$ such that A_i, $\mathsf{C} \subseteq \mathsf{R}$.

Theorem 12.2.2 shows that this is a consistent definition but $\Sigma \mathscr{A}$ is not necessarily defined for every C.S.T. $\mathscr{A}$. We shall say that a C.S.T. $\mathscr{A}$ is *standard* if there is an $A = \langle \mathsf{A}_i : \mathsf{C} \rangle$ such that for all $i \in \mathrm{C}‘\mathsf{C}$, $\mathsf{A}_i \subseteq \mathsf{R}$ and $\mathsf{C} \subseteq \mathsf{R}$. So a C.S.T. $\mathscr{A}$ is standard if, and only if, $\Sigma \mathscr{A}$ is defined.

12.2.4 THEOREM. $\mathscr{A}$ is standard if, and only if, for each $A \in \mathscr{A}$ there is a pair of partial recursive functions $p(i, x)$, $q(x)$ such that
(i) $q : \mathsf{C} \simeq \mathsf{D} \subseteq \mathsf{R}$ and
(ii) $\lambda x p(i, x) : \mathsf{A}_i \simeq \mathsf{B}_i \subseteq \mathsf{R}$;
i.e. p maps the A_i *uniformly* into R.

We now leave the reader to prove that having zeros in a S.O. makes no difference to its sum.
12.2.5 THEOREM. Let $A = \langle \mathsf{A}_i : \mathsf{C} \rangle$ and let $\mathsf{D} \subset \mathsf{C}$ and suppose that if $i \in \mathrm{C}‘\mathsf{C} - \mathrm{C}‘\mathsf{D}$ then $\mathsf{A}_i = \emptyset$. Then if $A' = \langle \mathsf{A}_i : \mathsf{D} \rangle$ and either side of (3) below is defined then so is the other and (3) holds:

$$\Sigma A \simeq \Sigma A' . \tag{3}$$

The only point to be borne in mind is that since only C is changed to D and we are always dealing with linear orderings embeddable in R this single change does not affect the uniform embedding of the A_i in R.

Let us write $\boldsymbol{O}_{\mathsf{A}} = \langle \emptyset : \mathsf{A} \rangle$ then we have:
12.2.6 COROLLARY. $\Sigma \boldsymbol{O}_{\mathsf{A}} = \emptyset$ for all $\mathsf{A} \subseteq \mathsf{R}$.
We shall also write $\mathscr{O} = \mathrm{CST}(\boldsymbol{O}_{\mathsf{A}})$ since, in general, the actual choice of A will be irrelevant.

There are two ways (at least!) of adding S.O.s; pointwise and (what we may call) ordinally. Pointwise addition seems not to be very useful so we only treat the other kind of addition.

12.2.7 DEFINITION. $\boldsymbol{A} + \boldsymbol{B} = \langle \mathsf{E}_i : \mathsf{C} \mathbin{\hat{+}} \mathsf{D} \rangle$ if

$$\boldsymbol{A} = \langle \mathsf{A}_i : \mathsf{C} \rangle, \ \boldsymbol{B} = \langle \mathsf{B}_i : \mathsf{D} \rangle,$$

where

$$\mathsf{E}_i = \mathsf{A}_i \quad \text{if} \quad i \in \mathrm{C}‘\mathsf{C}, \quad \mathsf{E}_i = \mathsf{B}_i \quad \text{if} \quad i \in \mathrm{C}‘\mathsf{D},$$

provided $\mathsf{C})(\mathsf{D}$. If $\mathsf{C})(\mathsf{D}$ we write also $\boldsymbol{A})(\boldsymbol{B}$ and say that $\boldsymbol{A}$ is *separable* from $\boldsymbol{B}$.

The pure theory of addition of C.S.T.s (i.e. that part of the theory not explicitly involving C.O.T.s) exactly parallels that of addition of C.O.T.s and since we shall be concerned with a finer treatment of sums of C.S.T.s we only give the definitions we require.

12.2.8 DEFINITION. $\mathscr{A}+\mathscr{B}=\mathrm{CST}(\boldsymbol{A}+\boldsymbol{B})$ for any $\boldsymbol{A}\in\mathscr{A}$, $\boldsymbol{B}\in\mathscr{B}$ such that $\mathsf{C})\,(\mathsf{D}$ (C, D as in definition 12.2.7).

That $\mathscr{A}+\mathscr{B}$ is well-defined follows from the following theorem.

12.2.9 THEOREM. If $\boldsymbol{A}\simeq\boldsymbol{A}'$, $\boldsymbol{B}\simeq\boldsymbol{B}'$, $\boldsymbol{A})\,(\boldsymbol{B}$ and $\boldsymbol{A}')\,(\boldsymbol{B}'$ then

$$\boldsymbol{A}+\boldsymbol{B}\simeq\boldsymbol{A}'+\boldsymbol{B}'.$$

PROOF. By hypothesis there exist

$$p, q: \boldsymbol{A}\simeq\boldsymbol{A}';\ r, s: \boldsymbol{B}\simeq\boldsymbol{B}'$$

and r.e. sets α, α', β, β' such that

$$\alpha\supseteq \mathsf{C}`\mathsf{C},\ \alpha'\supseteq\mathsf{C}`\mathsf{C}',\ \beta\supseteq\mathsf{C}`\mathsf{D},\ \beta'\supseteq\mathsf{C}`\mathsf{D}'$$

and $\alpha\cap\beta=\alpha'\cap\beta'=\emptyset$. Define

$$\begin{aligned} v(i) &= q(i) \quad \text{if} \quad i\in\alpha \quad \text{and} \quad q(i)\in\beta,\\ &= q'(i) \quad \text{if} \quad i\in\alpha' \quad \text{and} \quad q'(i)\in\beta',\\ &\quad \text{undefined otherwise} \end{aligned}$$

and

$$\begin{aligned} u(i, x) &= p(i, x) \quad \text{if} \quad i\in\alpha \quad \text{and} \quad q(i)\in\beta,\\ &= p'(i, x) \quad \text{if} \quad i\in\alpha' \quad \text{and} \quad q'(i)\in\beta',\\ &\quad \text{undefined otherwise.} \end{aligned}$$

Then the reader will readily complete the proof.

12.3 We can now obtain proofs of many analogues of the theorems we obtained for addition of C.O.T.s. We shall omit proof of most of these analogues: the essential difference is already contained in the proof of theorem 12.2.9. There the definition of u mirrors exactly the definition of v. We shall not need refinements until we deal with theorems which essentially involve (at least one of) the A_i. In this section we state some of the theorems on addition of C.S.T.s which do not involve Σ.

12.3.1 THEOREM. (i) $\mathscr{A}+\mathscr{O}=\mathscr{O}+\mathscr{A}=\mathscr{A}$,
(ii) $\mathscr{A}+\mathscr{B}=\mathscr{O}\Leftrightarrow\mathscr{A}=\mathscr{B}=\mathscr{O}$.

12.3.2 THEOREM. $+$ is associative, i.e.

$$\mathscr{A}+(\mathscr{B}+\mathscr{C})=\mathscr{A}+(\mathscr{B}+\mathscr{C}).$$

12.3.3 THEOREM. (cf. separation lemma). If $\mathscr{A}=\mathscr{B}+\mathscr{C}$ and $\boldsymbol{A}\in\mathscr{A}$ then there exist $\boldsymbol{B}\in\mathscr{B}$ and $\boldsymbol{C}\in\mathscr{C}$ such that $\boldsymbol{B})\,(\boldsymbol{C}$ and

$$\boldsymbol{A}=\boldsymbol{B}+\boldsymbol{C}.$$

12.3.4 THEOREM. (cf. directed refinement theorem). If $\mathscr{A}+\mathscr{C}=\mathscr{B}+\mathscr{D}$ then there is an $\mathscr{E}$ such that either

$$\mathscr{A}=\mathscr{B}+\mathscr{E} \quad\text{and}\quad \mathscr{E}+\mathscr{C}=\mathscr{D},$$

or

$$\mathscr{A}+\mathscr{E}=\mathscr{B} \quad\text{and}\quad \mathscr{C}=\mathscr{E}+\mathscr{D}.$$

Notation. $\mathscr{A}\cdot 0=\mathscr{O}$, $\mathscr{A}\cdot(n+1)=\mathscr{A}\cdot n+\mathscr{A}$.

12.3.5 THEOREM. (i) $\mathscr{A}\cdot(m+n)=\mathscr{A}\cdot m+\mathscr{A}\cdot n$,
(ii) $\mathscr{A}\cdot mn=(\mathscr{A}\cdot m)\cdot n$.

We conclude this section with a theorem which is not obtained in the same way as the above but is useful.

12.3.6 THEOREM. If $\boldsymbol{A}=\langle \mathsf{A}_i : \mathsf{B} \mathbin{\dot{+}} \mathsf{C}\rangle$, then

$$\Sigma\boldsymbol{A}\simeq\Sigma\boldsymbol{B}+\Sigma\boldsymbol{C}$$

where

$$\boldsymbol{B}=\langle \mathsf{A}_i : \mathsf{B}\rangle \quad\text{and}\quad \boldsymbol{C}=\langle \mathsf{A}_i : \mathsf{C}\rangle,$$

provided $\mathsf{B})\,(\mathsf{C}$.

12.4 We now consider finer versions of the separation lemma and the directed refinement theorem than those in 12.3.3,4. In TARSKI (1956), infinite sums are considered and although these do not hold in full generality we can make translations of them using Σ. Thus corresponding

to Tarski's postulate (III) (1956, p. 8) we have theorem 12.4.1, below.
Notation. If $\mathscr{A}$ is a C.S.T. and $A \in \mathscr{A}$ we write

$$\mathscr{A}_i = \mathrm{COT}(\mathsf{A}_i) \quad \text{where} \quad A = \langle \mathsf{A}_i : \mathsf{C} \rangle .$$

Clearly, $\mathscr{A}_i$ is well-defined if C is given. However, when we do not specify the C it will be clear either how C is to be chosen or that the assertion is purely existential and therefore independent of the particular choice of C from the various possibilities. This is the case in the next theorem and lemma.

12.4.1 THEOREM (generalized directed refinement theorem). If $\Sigma\mathscr{A} = B + C$ and $C \neq 0$, then there exist C.O.T.s D, E, C.S.T.s $\mathscr{A}^0$, $\mathscr{A}^1$ and an integer i such that either

$$\Sigma\mathscr{A}^0 + D = B,$$
$$E + \Sigma\mathscr{A}^1 = C \quad \text{and} \quad D + E = \mathscr{A}_i ,$$

or

$$\Sigma\mathscr{A}^0 = B \quad \text{and} \quad \Sigma\mathscr{A}^1 = C .$$

Before we prove this theorem we prove the following:

12.4.2 LEMMA (generalized separation lemma). If

$$\Sigma\mathscr{A} = B + C \quad \text{and} \quad A \in \mathscr{A} \ (A = \langle \mathsf{A}_i : \mathsf{D} \rangle).$$

Then there exist $\mathsf{B} \in B$, $\mathsf{C} \in C$, S.O.s A^0, A^1 and linear orderings E, F such that either

(i) $\Sigma A^0 \dotplus \mathsf{E} = \mathsf{B}$, $\mathsf{F} \dotplus \Sigma A^1 = \mathsf{C}$ and
$\mathsf{E} \dotplus \mathsf{F} = \mathsf{A}_i$ where $A = A^0 + \langle \mathsf{A}_i \rangle + A^1$ and
$\langle \mathsf{A}_i \rangle = \langle \mathsf{A}_i : \{\langle i,i \rangle\} \rangle$,

or (ii) $\Sigma A^0 = \mathsf{B}$, $\Sigma A^1 = \mathsf{C}$ and
$A^0 + A^1 = A$.

PROOF. By the separation lemma 2.3.1 given $\Sigma A \in \Sigma\mathscr{A}$ and $\Sigma A \simeq \mathsf{B}' + \mathsf{C}'$, where $\mathsf{B}' \in B$ and $\mathsf{C}' \in C$, there exist $\mathsf{B} \in B$ and $\mathsf{C} \in C$ such that

$$\mathsf{B})(\mathsf{C} \text{ and } \Sigma A = \mathsf{B} \dotplus \mathsf{C} .$$

Now $\mathrm{C}`\Sigma A = \{\hat{\imath}(a) : i \in \mathrm{C}`\mathsf{D} \,\&\, a \in \mathrm{C}`\mathsf{A}_i\}$.
Let

$$\mathsf{D}^0 = \{i : (\exists j)\,(\exists x)\,(\hat{\jmath}(x) \in \mathrm{C}`\mathsf{B} \ \&\ i \leq_{\mathsf{D}} j)\} .$$

Then, clearly $\mathsf{D}^0 = \mathsf{D}\,[\mathrm{D}^0$ is an initial segment of D. Let

$$\mathrm{D}^1 = \mathrm{C}\text{‘}\mathsf{D} - \mathrm{D}^0 \quad \text{and} \quad \mathsf{D}^1 = \mathsf{D}\,[\mathrm{D}^1 .$$

Now $\mathrm{D}^1 \supseteq \{i:(\exists j)(\exists x)(\hat{\jmath}(x) \in \mathrm{C}\text{‘}\mathsf{C}\;\&\; j \leq i\} = \mathrm{D}^2$, say. If $\mathrm{D}^1 = \mathrm{D}^2$ then $\mathsf{B})(\mathsf{C}$ implies $\mathsf{D}^0)(\mathsf{D}^1$. If $\mathrm{D}^1 \neq \mathrm{D}^2$ then $j \in \mathrm{D}^1 - \mathrm{D}^2$ implies $\mathsf{A}_j = \emptyset$ and in this case we redefine D^0, D^1 by $\mathsf{D}^0 = j)\mathsf{D}$ and $\mathsf{D}^1 = \mathsf{D}[j$ for such a j. That $\mathsf{D}^0)(\mathsf{D}^1$ then follows from lemma 2.1.2.

We now have two cases to consider.

(i) If $(\exists i)(\exists b)(\exists c)(\hat{\imath}(a) \in \mathrm{C}\text{‘}\mathsf{B}\;\&\;\hat{\imath}(b) \in \mathrm{C}\text{‘}\mathsf{C})$ and (ii) otherwise.

(ii) In this case we leave the reader to check that

$$\mathcal{A}^0 = \mathcal{A}\,[\mathrm{D}^0,\ \mathcal{A}^1 = \mathcal{A}\,[\mathrm{D}^1$$

suffice where

$$\begin{aligned} \mathcal{A}\,[\mathrm{G} &= \langle \mathsf{A}_i : \mathsf{D}\,[\mathrm{G}\rangle \\ &= \{\mathsf{A}_i : i \in \mathrm{C}\text{‘}\mathsf{D}\,[\mathrm{G}\}\,. \end{aligned}$$

In case (i) it is readily verified (cf. the proof of lemma 2.3.1) that there is a unique i for which

$$(\exists b)(\exists c)(\hat{\imath}(b) \in \mathrm{C}\text{‘}\mathsf{B}\;\&\;\hat{\imath}(c) \in \mathrm{C}\text{‘}\mathsf{C}).$$

Let

$$\begin{aligned} &\mathrm{D}_0 = \mathrm{D}^0 - \{i\},\ \mathrm{D}_1 = \mathrm{D}^1 - \{i\}, \\ &\mathcal{A}^0 = \mathcal{A}\,[\mathrm{D}_0,\ \mathcal{A}^1 = \mathcal{A}\,[\mathrm{D}_1\,. \end{aligned}$$

Then we have

$$\Sigma\mathcal{A}^0 \mathbin{\hat{+}} \mathsf{A}_i \mathbin{\hat{+}} \Sigma\mathcal{A}^1 = \mathsf{B} \mathbin{\hat{+}} \mathsf{C}\,.$$

Moreover, $\Sigma\mathcal{A}^0$ is an initial segment of B and $\Sigma\mathcal{A}^1$ is a final segment of C. Let

$$\mathsf{E} = \mathsf{B}\,[(\mathrm{C}\text{‘}\mathsf{B} - \mathrm{C}\text{‘}\Sigma\mathcal{A}^0)$$

and

$$\mathsf{F} = \mathsf{C}\,[(\mathrm{C}\text{‘}\mathsf{C} - \mathrm{C}\text{‘}\Sigma\mathcal{A}^1).$$

Then we leave the reader to check (again compare the proof of 2.3.1) that $\mathsf{E} \mathbin{\hat{+}} \mathsf{F} = \mathsf{A}_i$.

This completes the proof of the lemma and theorem 12.4.1 follows at once by taking C.S.T.s.

This lemma and theorem can be generalized to the case of $\Sigma\mathscr{A} = \Sigma\mathscr{B}$ but this (painful) exercise is left to the reader. The classical case is treated in TARSKI (1956) and the effective separability required is guaranteed by theorem 12.2.4 using heavily the fact that R is a recursive ordering.

12.4.3 COROLLARY. If $B \leq \Sigma\mathscr{A}$, and each (or equivalently, some) $A \in \mathscr{A}$ contains no zeros then there is an $\mathscr{A}_0$ such that for some $\mathscr{A}_1$

$$\mathscr{A} = \mathscr{A}_0 + \mathscr{A}_1,$$
$$B \leq \Sigma\mathscr{A}_0$$

and $\mathscr{A}_0$ is minimal with respect to these two properties.

PROOF. In case (i) of the theorem take

$$\mathscr{A}_0 = \mathscr{A}^0 + \text{C.S.T.}\langle \mathsf{A}_i : \{\langle i, i\rangle\}\rangle,$$

in case (ii) simply take $\mathscr{A}_0 = \mathscr{A}^0$.

12.5 From now until the end of this chapter we restrict our attention to well-orderings. The results on bounds apply almost without exception (*pace* the mention of classical ordinals) to quords and details of those results which can be readily adapted to the present framework may be found in CROSSLEY (1965).

12.5.1 DEFINITION. A co-ordinal A is said to be an *upper bound* for a collection of co-ordinals $\mathscr{A}$ if $B \in \mathscr{A}$ implies $B \leq A$. A is said to be the *least upper bound* if it is an upper bound and if C is any other upper bound then $A \leq C$. A is said to be a *minimal upper bound* if it is an upper bound and if C is any other upper bound then $C \nleq A$.

By the anti-symmetry of $\leq$ least upper bounds are unique if they exist at all. Also, it is clear that all least upper bounds are minimal, though the converse is certainly not true as we shall show. In fact we shall prove that all least upper bounds are trivial and determine when minimal upper bounds exist. Further we shall show that if one minimal bound exists then so do $2^{\aleph_0}$ mutually incomparable minimal bounds.

We can define lower bounds in the obvious way but we have no interesting results concerning them (cf. § 3 of CROSSLEY, 1965).

12.5.2 LEMMA. If a collection of co-ordinals $\mathscr{A}$ has an upper bound then any two members of the collection are comparable.

PROOF. Let A be an upper bound for $\mathscr{A}$, then $B, C \in \mathscr{A}$ implies $B \leq A$ and $C \leq A$ whence by theorem 4.2.8 B and C are comparable.

12.5.3 LEMMA. A collection of co-ordinals has a least upper bound (which is a co-ordinal) only if it is countable.

PROOF. Suppose $\mathscr{A}$ is a collection of co-ordinals with least upper bound A then $|A|$ is a countable ordinal, call it α. By lemma 12.5.2 if $B, C \in \mathscr{A}$ and $|B|=|C|=\beta\leq\alpha$ then by corollary 5.3.5, $B=C$. Hence $|\ |$ is a one-one function from $\mathscr{A}$ into the ordinals $\leq\alpha$. It follows that $\mathscr{A}$ is countable.

12.5.4 LEMMA. Suppose $p:\mathsf{C}\sim\mathsf{D}$ is a classical isomorphism, $\boldsymbol{A}=\langle \mathsf{A}_i:\mathsf{C}\rangle$, $\boldsymbol{B}=\langle \mathsf{B}_j:\mathsf{D}\rangle$, $\mathsf{A}_i=\mathsf{B}_{p(i)}$ for all $i\in \mathsf{C}`\mathsf{C}$ and for all i, $\min(\mathsf{A}_i)=0$, then

$$\Sigma\boldsymbol{A}\simeq\Sigma\boldsymbol{B} \quad \text{implies} \quad \mathsf{C}\simeq\mathsf{D}.$$

PROOF. Clearly the map defined by

$$q(\hat{\imath}(a)) = \widehat{p(i)}(a),$$

restricted to $\mathsf{C}`\Sigma\boldsymbol{A}$ is an isomorphism between $\Sigma\boldsymbol{A}$ and $\Sigma\boldsymbol{B}$. Hence, if

$$r:\Sigma\boldsymbol{A}\simeq\Sigma\boldsymbol{B},$$

then by theorem 8.3.4, r is an extension of q. Moreover, the map s defined by

$$s(i) = (r(\hat{\imath}(0)))_0,$$

is an extension of the (classical) isomorphism between C and D – this again by theorem 8.3.4 and the fact that $0=\min(\mathsf{A}_i)$ which means that s always yields a value for $i\in\mathsf{C}`\mathsf{C}$. Clearly s is partial recursive and one-one everywhere it is defined. Thus, as required, we have $s:\mathsf{C}\simeq\mathsf{D}$.

12.5.5 LEMMA. A strictly increasing (under $<$) sequence of co-ordinals of type ω has no least upper bound but has $2^{\aleph_0}$ minimal upper bounds.

PROOF. Let $\mathscr{A}=\{A_i\}_{i<\omega}$ be a strictly increasing sequence of co-ordinals, then by corollary 3.2.9 there exist unique $B_i(i<\omega)$ such that

$$A_0 = B_0,\ A_{n+1} = A_n + B_{n+1}.$$

Using the axiom of choice, choose $\mathsf{B}_i\in B_i$ so that $\mathsf{B}_i\subseteq\mathsf{R}$. Then, since the sequence $\{A_i\}$ is strictly increasing, no B_i is empty. Let

$$\boldsymbol{B}^{(\mathsf{W})} = \langle \mathsf{B}_i:\mathsf{W}\rangle \quad\text{and let}\quad \boldsymbol{B}^{(\mathsf{U})} = \langle \mathsf{D}_i:\mathsf{U}\rangle, \quad\text{where}\quad \mathsf{D}_{|i|_\mathsf{U}} = \mathsf{B}_i$$

and U is any well-ordering of type ω not recursively isomorphic to W.

By theorem 5.2.4 U may be chosen in $2^{\aleph_0}$ different ways. Finally let

$$B^{(W)} = \mathrm{COT}\,\boldsymbol{\Sigma}(\boldsymbol{B}^{(\mathsf{W})}),\ B^{(U)} = \mathrm{COT}\,\boldsymbol{\Sigma}(\boldsymbol{B}^{(\mathsf{U})}).$$

We claim that $B^{(W)}$ and all the $B^{(U)}$ are minimal upper bounds for $\mathscr{A}$.

If $A \in \mathscr{A}$ then $A = A_n$ for some $n < \omega$ and hence by theorem 12.3.6,

$$\begin{aligned} A_n = \sum_{i=1}^{n} B_i &\leq \sum_{i=1}^{n} B_i + \mathrm{COT}(\boldsymbol{\Sigma}\langle \mathsf{B}_i : i > n\rangle), \\ &= B^{(W)}. \end{aligned} \tag{4}$$

Similarly, one shows $B^{(U)}$ is an upper bound for $\mathscr{A}$. By lemma 12.5.4, all the $B^{(U)}$ are incomparable with $B^{(W)}$ and with each other and it only remains to show that if $C < B^{(U)}$ then $C < A_n$ for some n. But this is obvious from the classical case (though it can be proved by using the fact that $\mathsf{C}`\boldsymbol{\Sigma}(\boldsymbol{B}^{(\mathsf{U})}) - \mathsf{C}`\mathsf{C}$ is non-empty for any $\mathsf{C} \in C$).

Summing up and dealing with peripheral cases we obtain:

12.5.6 THEOREM. Let $\mathscr{A}$ be a collection of co-ordinals and let $\alpha = \sup_{A \in \mathscr{A}} |A|$ (here we permit $\alpha = \Omega$, the first uncountable ordinal). Then

(i) $\mathscr{A}$ has a least upper bound if, and only if, $\mathscr{A}$ (is countable and) has a maximum.

(ii) If $\mathscr{A}$ is countable, all the $A \in \mathscr{A}$ are comparable and $\mathscr{A}$ has no maximum then $\mathscr{A}$ has $2^{\aleph_0}$ mutually incomparable minimal upper bounds (but no least upper bound).

(iii) If $\mathscr{A}$ is uncountable then $\mathscr{A}$ has no upper bound.

PROOF. (i) follows at once from the preceding lemmata.

(ii) Assume the hypotheses. Since $\mathscr{A}$ is countable, α is countable hence there is a strictly increasing ω-sequence $\{C_i\}_{i<\omega}$ co-final (w.r.t. $\leq$) with $\mathscr{A}$. The result now follows easily from the preceding lemma using the C_i for the A_i of that lemma.

(iii) If α is uncountable there is nothing to prove. If α is countable then there exists a countable ordinal $\beta \leq \alpha$ for which there are uncountably many (distinct) co-ordinals $B \in \mathscr{A}$ such that $|B| = \beta$. But if $\mathscr{A}$ had an upper bound this would contradict theorem 5.3.4. This completes the proof.

12.5.7 LEMMA. If A is a minimal upper bound for $\mathscr{A}$ then

$$|A| = \sup_{B \in \mathscr{A}} |B|.$$

PROOF. Assume the hypothesis and let $\sup_{B \in \mathscr{A}} |B| = \alpha$. By corollary 5.3.6 there is a unique $C \leq A$ such that $|C| = \alpha$. The reader will now easily verify (using theorem 4.2.8) that C is also an upper bound for $\mathscr{A}$ and hence $C = A$ (since C, A are comparable).

12.6 Now we deal with the question of paths $\mathscr{P}$ (cf. chapter 5) which go "right through" the classical ordinals in the following sense.

12.6.1 DEFINITION. A *long path* (of co-ordinals) is a set of co-ordinals $\mathscr{A}$ linearly (& so well-)ordered by $\leq$ such that for each (countable) ordinal α there exists (a unique) $A \in \mathscr{A}$ with $|A| = \alpha$.

12.6.2 THEOREM. There exist 2^c long paths.

PROOF. We prove by transfinite induction on the classical ordinals that for each ordinal α there is a co-ordinal A_α such that (i) $|A_\alpha| = \alpha$ and (ii) A_α is comparable with all A_β for $\beta < \alpha$. If $\alpha = 0$ the assertion is trivial. By theorem 5.1.6 we can take $A_{\beta+1} = A_\beta + 1$. Now suppose that α is a limit number. By the induction hypothesis all A_β for $\beta < \alpha$ are comparable, hence by theorem 12.5.6 $\{A_\beta\}_{\beta<\alpha}$ has $2^{\aleph_0}$ minimal upper bounds. Let A_α be any one of these. Clearly $\{A_\beta\}_{\beta<\alpha}$ has the required properties. Now different choices of A_α can be made at each limit number and the actual choice can be made in $c = 2^{\aleph_0}$ ways. All these choices give different paths (since two paths are distinct if they are different at one point), and hence there are $c^c = 2^c$ such paths. This completes the proof.

As an immediate corollary we obtain the existence of long closed paths.

12.6.3 THEOREM. There exist 2^c long paths $\mathscr{P}$ such that $\langle \mathscr{P}, +, \cdot, \exp \rangle$ is isomorphic to $\langle \Omega, \oplus, \odot, exp \rangle$ by $|\ |$.

PROOF. Let $\mathscr{Q}$ be any long path and let $\mathscr{P} = \mathrm{E}(\mathscr{Q}) = \{B : B \leq \mathrm{E}(A)$ for some $A \in \mathscr{Q}\}$. By theorem 11.2.11 and lemma 11.1.5 we obtain the required result.

12.7 In this section we prove a few theorems about principal numbers including Aczel's unpublished result that there exist principal numbers for addition, and for multiplication not of the forms W^A, W^{W^A}, respectively.

12.7.1 THEOREM. If $\mathscr{P}=\{P_i\}_{i=0}^{\infty}$ is a strictly increasing sequence of principal numbers for addition then there exist $2^{\aleph_0}$ principal numbers for addition which are minimal upper bounds for $\mathscr{P}$.

PROOF. Let $P^{(U)}$ be constructed exactly as in the proof of lemma 12.5.5 (with "P" for "A" everywhere) then for each co-ordinal U of classical ordinal ω we obtain one minimal upper bound $P^{(U)}$. We now show that

$$P^{(U)} \in \mathscr{H}(+).$$

By (4)

$$\begin{aligned} &P_n \leq P^{(U)} \quad \text{for all } n \text{ and also} \\ &B < P^{(U)} \Rightarrow (\exists n)\,(B < P_n), \end{aligned} \tag{5}$$

as is easily verified. Hence, using the fact that $P_n \in \mathscr{H}(+)$,

$$\begin{aligned} B < P^{(U)} \Rightarrow B + P^{(U)} &= B + P_n + (P^{(U)} - P_n) \\ &= P_n + (P^{(U)} - P_n) \\ &= P^{(U)}, \end{aligned}$$

where n is the number given by (5). We conclude that $P^{(U)} \in \mathscr{H}(+)$ for all U with $|U| = \omega$. This completes the proof.

12.7.2 LEMMA. There are $2^{\aleph_0}$ co-ordinals U (of type ω) such that $U \subseteq W$ and $W \neq U$, similarly for well-orderings.

PROOF. The former assertion follows from the latter since C.O.T.s are always countable classes of linear orderings. C'W has $2^{\aleph_0}$ infinite, and therefore cofinal, subsets. Let U be such a set with the ordering induced by W.

12.7.3 LEMMA. If $B = W \cdot A$ and $\mathsf{B} \in B$ then there is a function $\lambda x s(x)$ such that in B $|s(x)| = |x| + 1$, which has a one-one partial recursive extension.

PROOF. Clearly it suffices to consider $\mathsf{B} = \mathsf{W} \cdot \mathsf{A}$, but then $s(x) = j(k(x)+1, l(x))$ is such a function.

12.7.4 THEOREM. Let $\mathscr{P}=\{P_i\}_{i=0}^{\infty}$ be a strictly increasing sequence of principal numbers for addition; then there exist $2^{\aleph_0}$ principal numbers for addition which are minimal upper bounds for $\mathscr{P}$ and which are not of the form W^A.

PROOF. We repeat the construction of theorem 12.7.1. but with one change. We choose the B_i's (see the proof of lemma 12.5.5) so that $\min(\mathsf{B}_i)=0$ for all i and we set $\min(\mathsf{B}_i[(\mathsf{C}`\mathsf{B}_i-\{0\}])=u_i$. Clearly this can be done whilst still ensuring $\mathsf{B}_i\subseteq\mathsf{R}$; for example, we can take B_i' as in the proof of lemma 12.5.5 and then take $\mathsf{B}_i=\widehat{(u_i+1)}\ \mathsf{B}_i'$ before arranging the first two elements as above. Now we construct $P^{(U)}$ as before where $\mathsf{U}=\{\langle u_i, u_j\rangle : i\leq j\}$ is a well-ordering satisfying the conclusion of lemma 12.7.2. Suppose $P^{(U)}=W\cdot A$ for some A; then by lemma 12.7.3 there is a one-one partial recursive function s such that $|s(x)|=|x|+1$ in $\mathsf{P}^{(\mathsf{U})}$. In particular, therefore we have

$$sj(i,0)=j(i,u_i),$$

whence $\lambda i u_i$ is a partial recursive function, defined on the whole of the natural numbers and is one-one. It follows that $u:\mathsf{W}\simeq\mathsf{U}$ which is a contradiction. This completes the proof.

12.7.5 Corollary. If $\mathscr{A}=\{W^{W^{A_i}}\}_{i=0}^{\infty}$ is a strictly increasing sequence of principal numbers for multiplication then there exist $2^{\aleph_0}$ minimal upper bounds for $\mathscr{A}$ which are principal numbers for multiplication not of the form W^{W^A}.

Proof. By theorem 12.7.4, there are $2^{\aleph_0}$ minimal upper bounds A for $\{W^{A_i}\}_{i=0}^{\infty}$ which are principal numbers for *addition* and are not of the form W^B. By theorem 10.2.7, W^A is a principal number for multiplication. We leave the reader to check that each such W^A is a minimal upper bound for $\mathscr{A}$. Finally, if W^A were of the form W^{W^B} we should have $A=W^B$ by theorem 8.5.3 which contradicts the construction of A. This completes the proof.

PART TWO

CHAPTER 13

ISOLS[131]

13.1 So far we have been studying the properties of equivalence classes of ordered sets under one-one, partial recursive, order preserving maps and have found both similarities with and differences from the classical theory. One of the initial stimuli for this study was DEKKER and MYHILL's (1960) treatment of a recursive analogue of cardinal number theory and we now include a brief review of their (and others') results insofar as these parallel ours in the remaining chapters of this monograph.[132]

None of the definitions and results in this chapter are due to the present author. Throughout this chapter we shall give references to the order analogues in this monograph and when the proofs required here differ only slightly or not at all from those in the order analogue cases we shall omit them leaving the reader to look them up and make the necessary amendments. Since some of these will be references to the next two chapters the reader who is not familiar with isols and R.E.T.s may care to regard the omitted proofs as exercises.

13.1.1 DEFINITION (cf. definition 1.1.4). A set A is said to be *recursively equivalent* to (a set) B if there is a one-one, partial recursive map p such that

$$\mathrm{A} \subseteq \delta p \quad \text{and} \quad p(\mathrm{A}) = \mathrm{B}.$$

We write $p\colon \mathrm{A} \simeq \mathrm{B}$ or $\mathrm{A} \simeq \mathrm{B}$ in this case.[133]

As in the order case (see the remarks following definition 1.1.4) recursive equivalence is an equivalence relation. We set

$$A = \mathrm{RET}(\mathrm{A}) = \{\mathrm{B} : \mathrm{B} \simeq A\}$$

and call A a *recursive equivalence type* (R.E.T.). So the R.E.T.s are the "cardinal" analogues of C.O.T.s.

Definition 13.1.1 is not exactly imitated by definition 1.1.4, the precise analogue would require $p : \mathrm{A}' \simeq \mathrm{B}'$ (in the above sense) where $\mathrm{A} \subseteq \mathrm{A}'$, $\mathrm{B} \subseteq \mathrm{B}'$ and A′, B′ are r.e. However, since δp and ρp are r.e. it follows at once that the two definitions of recursive equivalence coincide. Moreover, since $+, \cdot$ and exponentiation are determined by partial recursive functionals all theorems concerning these operations are unaffected by the different definitions.

13.1.2 Definition. A set A is said to be *isolated* if it is finite or immune.

13.1.3 Theorem (cf. theorem 15.1.1). If $\mathrm{A} \simeq \mathrm{B}$ and

A is $\begin{cases} \text{finite} \\ \text{immune} \end{cases}$ then B is $\begin{cases} \text{finite} \\ \text{immune} \end{cases}$

13.1.4 Theorem (cf. theorem 15.1.2). The following statements are equivalent.

(i) A contains an infinite r.e. subset,

(ii) A contains an infinite recursive subset,

(iii) A is not isolated.

Proof. (i)⇒(ii). Every infinite r.e. set contains an infinite recursive subset as follows: Let S be an infinite r.e. set and let a be some element of S. Set

$$g(0) = a,$$
$$g(n+1) = \nu_x\{x \in \mathrm{S} \;\&\; x > g(n)\},$$

where ν_x is Kleene's indefinite description operator. Then g is clearly partial recursive and everywhere defined since S is infinite. Therefore g is recursive and enumerates a subset T of S in order of magnitude so T is a recursive set.

The other implications are trivial.

13.1.5 Definition. We set $R = \mathrm{RET}(\mathcal{N})$, $n = \mathrm{RET}(\{0, 1, \ldots, n-1\})$ for $n \in \mathcal{N}$.

So R is the "cardinal" of W (the C.O.T. of the standard ω-ordering).

We can define (cardinal) addition of R.E.T.s again using separability conditions to ensure a proper definition.

13.1.6 DEFINITION (cf. definition 2.2.3). $A+B=\mathrm{RET}(\mathrm{A}\cup\mathrm{B})$ for any $\mathrm{A}\in A$, $\mathrm{B}\in B$ such that A) (B.

As in chapter 2 we can show that any two R.E.T.s have separable representatives.

13.1.7 THEOREM (cf lemma 2.2.1). If $\mathrm{A}_0 \simeq \mathrm{A}_1$, $\mathrm{B}_0 \simeq \mathrm{B}_1$, A_0) (B_0 and A_1) (B_1, then

$$\mathrm{A}_0 \cup \mathrm{B}_0 \simeq \mathrm{A}_1 \cup \mathrm{B}_1 .$$

13.1.8 THEOREM (cf. theorems 2.2.4,5).

(i) $+$(for R.E.T.s) is associative and commutative,

(ii) $A+0=A$.

We also have a separation lemma and refinement theorem but although the separation lemma is virtually identical with that for C.O.T.s *viz.* if $\mathrm{A}\in A$ and $A=B+C$ then $(\exists \mathrm{B}\in B)\,(\exists \mathrm{C}\in C)\,(\mathrm{A}=\mathrm{B}\cup\mathrm{C}\,\&\,\mathrm{B})$ (C), since we have no order properties the refinement theorem becomes:

13.1.9 THEOREM (cf. theorem 2.3.2). If $A+B=C+D$, then there exist $E_1, \ldots, E_4$ such that

$$A = E_1 + E_2, \quad B = E_3 + E_4,$$
$$C = E_1 + E_3, \quad D = E_2 + E_4.$$

PROOF. By the separation lemma if $\mathrm{A}\in A$, $\mathrm{B}\in B$ and A) (B, there exist $\mathrm{C}\in C$, $\mathrm{D}\in D$ such that $\mathrm{A}\cup\mathrm{B}=\mathrm{C}\cup\mathrm{D}$. Let $E_1=\mathrm{A}\cap\mathrm{C}$, $\mathrm{E}_2=\mathrm{A}\cap\mathrm{D}$, $\mathrm{E}_3=\mathrm{B}\cap\mathrm{C}$ and $\mathrm{E}_4=\mathrm{B}\cap\mathrm{D}$. Then we leave the reader to check E_i)(E_j if $i\neq j$, using A) (B and C) (D, and to complete the proof.

13.2.

13.2.1 DEFINITION. $A\cdot B=\mathrm{RET}(j(\mathrm{A}, \mathrm{B}))$ for any $\mathrm{A}\in A$, $\mathrm{B}\in B$.

13.2.2 THEOREM (cf. theorem 6.1.3). If $\mathrm{A}_0 \simeq \mathrm{A}_1$ and $\mathrm{B}_0 \simeq \mathrm{B}_1$, then

$$j(\mathrm{A}_0, \mathrm{B}_0) \simeq j(\mathrm{A}_1, \mathrm{B}_1).$$

This theorem justifies definition 13.2.1.

13.2.3 THEOREM (cf. theorems 6.1.8 and 6.2.1). Multiplication ($\cdot$) of R.E.T.s is associative and commutative and is distributive over addition.

13.2.4 DEFINITION. $A \leq B \Leftrightarrow (\exists C)(A + C = B)$.
$A \subseteq B \Leftrightarrow (\exists \mathrm{A} \in A)(\exists \mathrm{B} \in B)(\mathrm{A} \subseteq \mathrm{B})$.

13.2.5 THEOREM (cf. lemma 6.3.1). $B + A = A$ if, and only if, $B \cdot R \leq A$.

The proof of this theorem is rather more complicated than that of lemma 6.3.1 and we only indicate the main lines of the proof. Full details may be found in DEKKER and MYHILL (1960, pp. 75–77). Suppose we are given

$$f : \mathrm{B} \cup \mathrm{A} \simeq \mathrm{A} \quad \text{and} \quad \mathrm{B})(\mathrm{A}.$$

Then there exist disjoint r.e. sets $\mathrm{B}_1 \supseteq \mathrm{B}$ and $\mathrm{A}_1 \supseteq \mathrm{A}$. Let A^* be the r.e. set of natural numbers $x \in \mathrm{A}_1$ such that $x > f^{-1}(x)$ and for some r

$$x \in \mathrm{A}_1, f^{-1}(x) \in \mathrm{A}_1, \ldots, f^{-r+1}(x) \in \mathrm{A}_1, f^{-r}(x) \in \mathrm{B}_1.$$

In order to ensure separability we use the fact that the natural numbers are well-ordered so the s below is finitely determinable and we set $p(x) = j(s, f^{-r}(x))$ where s is the largest number of terms which can occur in a (finite) sequence (with $k_1 < k_2 < \cdots < r$)

$$x > f^{-k_1}(x) > \cdots > f^{-k_{s-2}}(x) > f^{-r}(x).$$

The rest of this part of the proof consists in checking that for every $y \in \mathrm{B}$ and every s there is an $x \in \mathrm{A}$ such that $p(x) = j(s, y)$ and also that $\mathrm{A}^*)((\mathrm{B} \cup \mathrm{A}) - \mathrm{A}^*$. In fact, since $f^n(y)$ is defined for all n if $y \in \mathrm{B}$ and since f is one-one x, s always exist it is clear that $j(s, x) \in R \cdot B$. Separability is also easy to establish since the set of x's which are mapped is recursive in A^*.

The converse follows easily from $1 + R = R$ and theorem 13.2.3 (cf. theorem 5.2.6.(i)).

13.2.6 THEOREM (cf. theorem 8.4.1). If $B \leq C$, then $AB \leq AC$.

13.2.7 THEOREM (cf. theorem 4.2.8). $\leq$ is reflexive, antisymmetric and transitive (on R.E.T.s).

PROOF. Reflexivity and transitivity are clear. If $B \leq A$ and $A \leq B$ then for some C, D, $B+C=A$ and $A+D=B$ so $A+(D+C)=A$ whence by theorem 13.2.5, $(D+C)\cdot R \leq A$. Now $D \leq D+C$ hence by theorem 13.2.6 $D\cdot R \leq (D+C)\cdot R \leq A$ and therefore by theorem 13.2.5,

$$A = D + A = B.$$

13.2.8 THEOREM (cf. theorems 13.1.4 and 15.1.2). The following statements are equivalent for an R.E.T. A:

(i) $\mathrm{A} \in A$ implies A is isolated,

(ii) $R \nleq A$,

(iii) $B \subseteq A \Rightarrow B+1 \neq B$,

(iv) $B \subseteq A$ & $B+C=B+D \Rightarrow C=D$,

(v) $B \subseteq A$ and $\mathrm{B} \in B \Rightarrow \mathrm{B}$ is isolated.

PROOF. Clearly the negation of (ii) implies the negation of (i).

(iii)$\Leftrightarrow$(i) follows easily from theorem 13.2.5.

(iv)$\Leftrightarrow$(iii) is trivial.

(iii)$\Leftrightarrow$(ii) follows from theorem 13.2.5.

(i)$\Rightarrow$(v) is trivial so there remains (v)$\Rightarrow$(iv).

We shall merely sketch the proof, full details being in DEKKER and MYHILL (1960) pp. 89–91.

Suppose $B+C=B+D$ then there exist $\mathrm{B} \in B$, etc. and a one-one, partial recursive map

$$p : \mathrm{B} \cup \mathrm{C} \simeq \mathrm{B} \cup \mathrm{D},$$

where B) (C, B) (D and we may assume C) (D. Consequently there exist disjoint r.e. sets $\mathrm{B}_1 \supseteq \mathrm{B}$, $\mathrm{C}_1 \supseteq \mathrm{C}$ and $\mathrm{D}_1 \supseteq \mathrm{D}$. Restrict domain of p to $\mathrm{B}_1 \cup \mathrm{C}_1$ and range of p to $\mathrm{B}_1 \cup \mathrm{D}_1$. We leave the reader to fill in the details to show that the map, q, defined below is such that $q : \mathrm{C} \simeq \mathrm{D}$. $q(x)$ is defined if, and only if, there is a finite sequence.

$$x,\ p(x), \ldots, p^r(x), \tag{1}$$

such that $x \in \mathrm{C}_1$, $p(x), \ldots, p^{r-1}(x) \in \mathrm{B}_1$ and $p^r(x) \in \mathrm{D}_1$. We then set $q(x)= p^r(x)$. $p^k(x)$ cannot always be in B if $x \in \mathrm{C}$ for then, since p is one-one, we should have an infinite r.e. subset of B contradicting (v). Moreover, we can effectively decide which of B_1, C_1, D_1 x is in if $x \in \mathrm{B}_1 \cup \mathrm{C}_1 \cup \mathrm{D}_1$ so the finite sequence (1) can be effectively generated and it follows that q is partial recursive. q obviously has an inverse given by considering sequences like (1) with p^{-k} for p^k, C_1 for D_1 and D_1 for C_1.

The above theorem illustrates the significance of the R.E.T.s of isolated sets.

13.2.9 DEFINITION. An R.E.T. A is said to be an *isol* if it satisfies (any of) the conditions in theorem 13.2.8.

13.2.10 COROLLARY. An isol contains a recursive set if, and only if, it is finite.

There are in general $\aleph_0$ partial recursive, one-one maps from an infinite isolated set onto itself – just exchange two points in the set – but we shall show in the next chapters (theorems 14.1.4 and 15.1.3) that if the set is ordered and the map required to be order preserving then there is only one of these maps. However, we cannot map isolated sets into proper subsets of themselves so isols are effective analogues of Dedekind finite numbers.

13.2.11 LEMMA. A is isolated if, and only if, $\mathrm{B}\subset\mathrm{A}\Rightarrow\mathrm{B}\not\simeq\mathrm{A}$.

PROOF. If A is infinite and not isolated, A contains an infinite recursive set $\mathrm{B}=\{b\}\cup\mathrm{B}_1$, $b\notin\mathrm{B}_1$. But (since $1+R=R$) there is a one-one, partial recursive map $p:\mathrm{B}\simeq\mathrm{B}_1$ so set $f(x)=p(x)$ if $x\in\mathrm{B}$, $f(x)=x$ otherwise then clearly $f:\mathrm{A}\simeq\mathrm{A}-\{b\}$.

Conversely, if there exists $\mathrm{B}\subset\mathrm{A}$ with $p:\mathrm{A}\simeq\mathrm{B}$ then if $a\in\mathrm{A}-\mathrm{B}$, $p(a)\neq a$ and $p^n(a)\neq p^m(a)$ for $n\neq m$. But $p^n(a)\in\mathrm{A}$ for all n so $\{p^n(n):n=0,1,\dots\}$ is an infinite r.e. subset of A. If A is finite the lemma is trivially true.

13.2.12 THEOREM (cf. theorem 15.2.2). (i) $A\leq B$ implies $A\subseteq B$,

(ii) Every infinite isol has $\aleph_0$ $\leq$-predecessors and $2^{\aleph_0}$ $\subseteq$-predecessors.

(iii) $A\subseteq B$ does not imply $A\leq B$.

PROOF. (i) is obvious.

(ii) $B\leq A$ only if for $\mathrm{A}\in A$ there exists $\mathrm{B}\in B$ and a C such that B) (C and $\mathrm{B}\cup\mathrm{C}=\mathrm{A}$. Since each r.e. set only separates off at most one B, A has at most $\aleph_0$ $\leq$-predecessors. $n\leq A$ for all n so there are exactly $\aleph_0$ $\leq$-predecessors. $A \in A$ implies A has $2^{\aleph_0}$ distinct subsets and these are divided into R.E.T.s containing at most $\aleph_0$ elements each.

(iii) follows trivially from (ii).

13.3.

13.3.1 THEOREM (cf. theorem 15.3.1). If A, B are isols then so too are $A+B$, $A \cdot B$.

The proof is virtually given as the proof of theorem 15.3.1 below.

We can define exponentiation as follows: Since A, B are sets of integers we may regard them as embedded in Seq by considering $A' = \{2^{a+2} : a \in A\}$ and then we can map min[A′] to 0 to obtain A_1 without changing the C.O.T. or R.E.T. of A. Next we define

$$A^B = C`[A_1] \exp [B_1],\ A^B = \mathrm{RET}(A^B).$$

However, we shall not prove any theorems about exponentiation of R.E.T.s but shall leave that to the ardent reader. As in theorem 15.3.1 below we can show that isols are closed under exponentiation.

We now turn to cancellation laws for isols. We have already pointed out that for isols $A+B=A+C$ implies $B=C$. In general we have results like $f(X, Y)=f(X, Z)$ implies $Y=Z$ for isols X, Y, Z (possibly with side conditions). However, the result of replacing "=" by "≤" is generally to make the implications false though we shall see that for losols the implications still go through. We shall not give proofs for these are, in general, of three kinds: (1) (for theorem 13.3.2 below) almost the same proof as DEKKER and MYHILL (1960) yields the result for isols and losols, (2) (for theorem 13.3.3, below) quite complicated counter-examples are required for isols but proofs as in (1) will work for losols, (3) (for theorem 13.3.4 below) special (and differing) techniques are required for the two theories.

In the theorems below all R.E.T.s are assumed to be isols unless otherwise stated.

13.3.2 THEOREM (i) (cf. theorem 16.1.2). If $A \neq 0$ and $AB=AC$, then $B=C$.

(ii) (cf. theorem 16.3.2). If $A \geq 2$ and $A^B = A^C$, then $B = C$.
(iii) (cf. theorem 16.3.2). If $A \neq 0$ & $B^A = C^A$, then $B = C$.

13.3.3 THEOREM. The following statements are *false*.
(i) (contrast theorem 16.2.2). If $A \neq 0$ and $AB < AC$, then $B \leq C$.
(ii) (contrast theorem 16.3.2). If $A \neq 0$ and $A^B < A^C$, then $B \leq C$.
Nerode has also shown (unpublished) that for isols
$A \neq 0$ and $B^A < C^A$ do not imply $B \leq C$.

13.3.4 THEOREM (cf. theorem 2.4.11 and appendix A). (i) (FRIEDBERG, 1961). If A, B are arbitrary R.E.T.s and $n \neq 0$, then

$$A \cdot n = B \cdot n \Rightarrow A = B .$$

(ii) (ELLENTUCK, 1963). There exists an isol X and R.E.T.s A, B such that $X \neq 0$, $XA = XB$ and $A \neq B$.

Finally, NERODE (1961) has shown that strong metatheorems go through for isols. We state only a very weak version which parallels our theorem 16.4.2.

13.3.5 THEOREM (NERODE, 1961). If $P(X)$ is a function of X alone constructed by finite composition from functions of the forms

$$X + A,\ X \cdot A_1,\ (1 + X)^{A_1},\ A_2^X,$$

(where $A_1 \geq 1$ and $A_2 \geq 2$) then

$$P(X) = P(Y) \Rightarrow X = Y$$

whenever X, Y are isols and all the parameters A_i are isols.

CHAPTER 14

QUASI-FINITENESS

14.1 As we noted in the previous chapter the theory of recursive equivalence types has tended to be most interesting when R.E.T.s of isolated sets have been considered; that is, sets with no infinite r.e. subset. The question naturally arises as to what are the order analogues of isols? Originally, Kreisel proposed quords to the author, but PARIKH (1962, 1966) showed that quords are not closed under exponentiation and it was not until the Leicester Logic Colloquium in 1965 that Nerode and the author came to the conclusion that what are the most natural analogues are *losols*. We shall introduce these in the next chapter. However, there is a larger class of C.O.T.s which the author considered in his thesis (1963) and the appendix to (1965) and we now take a brief look at the standard C.O.T.s of this type.

Classically, finite ordered sets may be defined as linearly ordered sets which contain no descending or ascending chains. So we now consider (cf. § 3.1) linearly ordered sets with no *recursive* descending or ascending chains. As in chapter 3, for sets embeddable in R by a recursive isomorphism, this condition is equivalent to every non-empty subset having a minimum and a maximum element.

We recall that $\mathsf{A}^* = \{\langle y, x\rangle : \langle x, y\rangle \in \mathsf{A}\}$, $A^* = \mathrm{COT}(\mathsf{A}^*)$ for any $\mathsf{A} \in A$. A^* is the converse of A and A^* the converse of A.

14.1.1 DEFINITION. A linear ordering A is said to be *quasi-finite* if A, A^* are quasi-well-orderings. A C.O.T. A is said to be *quasi-finite* if A, A^* are quords.

14.1.2 THEOREM. (i) If A is quasi-finite and $\mathsf{A} \in A$, then A is quasi-finite.

(ii) If some $\mathsf{A} \in A$ is quasi-finite, then every $\mathsf{A} \in A$ is quasi-finite.

PROOF. Immediate from theorem 3.2.2.

Notation. We write $B \subseteq A$ if, for some $\mathsf{A} \in A$ there is a $\mathsf{B} \in B$ with $\mathsf{B} \subseteq \mathsf{A}$. (Since $B \subseteq A$ in the set-theoretic sense implies $B = A$ because A, B are equivalence classes, this notation will not cause confusion.) We also write $B \subset A$ if $B \subseteq A$ and $B \neq A$.

14.1.3 THEOREM. The following statements are equivalent for a C.O.T. A.

(i) A is quasi-finite,

(ii) $B \subseteq A$ implies B is quasi-finite,

(iii) $W \nsubseteq A \,\&\, W^* \nsubseteq A$,

(iv) $B \subseteq A$ implies $B + C = B + D \Rightarrow C = D . \&. C + B = D + B \Rightarrow C = D$,

(v) $B \subseteq A \Rightarrow B + 1 \neq B \,\&\, 1 + B \neq B$.

PROOF. (i)$\Leftrightarrow$(ii), (iv)$\Rightarrow$(v). Left to the reader.

(ii)$\Rightarrow$(iii). $\mathsf{W} = \{\langle i, j \rangle : i \leq j\} \in W$.

If $W \subseteq A$ and $\mathsf{A} \in A$, then there exists a recursive isomorphism

$$p : \mathsf{W} \simeq \mathsf{W}' \subseteq \mathsf{A} .$$

But then $\{p(i)\}_{i=0}^{\infty}$ is a recursive ascending chain in A so A^* is not a quord. Similarly if $W^* \subseteq A$.

(iii)$\Leftrightarrow$(v) follows from lemma 6.3.1.

(iii)$\Rightarrow$(i). Suppose A is not quasi-finite. Let $\mathsf{A} \in A$ then A or A^* is not a quasi-well-ordering. Suppose $\{a_i\}_{i=0}^{\infty}$ is a recursive descending chain in A. Then $\lambda i a_i$ is a one-one partial recursive function (which is everywhere defined) such that $i < j$ if, and only if, $a_j <_{\mathsf{A}} a_i$. Hence $\{\langle a_i, a_j \rangle : i \leq j\} \in W^*$. Similarly if $\{a_i\}_{i=0}^{\infty}$ is a recursive ascending chain in A.

(i)$\Rightarrow$(iv) follows from definition 14.1.1, corollary 3.2.9 and theorem 2.4.4.

14.1.4 THEOREM. If A is quasi-finite then $p : \mathsf{A} \simeq \mathsf{B} \subseteq \mathsf{A}$ implies p extends the identity on $C'\mathsf{A}$.

PROOF. We may assume without loss of generality that $\mathsf{B} \subseteq \mathsf{A} \subseteq \mathsf{R}$. Suppose $p(a) \neq a$ for some $a \in C'\mathsf{A}$, then $p(a) \prec a$ or $a \prec p(a)$. Since p is order preserving and $\delta p \supseteq C'\mathsf{A}$ it follows that $\{p^n(a)\}$ is a recursive descending or ascending chain in A.

We do not know whether the converse is true – we suspect it is not – but we have the following weak converse.

14.1.5 THEOREM. If A is not quasi-finite then there exists $\mathsf{B} \subseteq \mathsf{A}$ and $\mathsf{C} \subset \mathsf{B}$ such that $\mathsf{B} \simeq \mathsf{C}$.

PROOF. Immediate from theorem 14.1.3.(v) and lemma 6.3.1.

14.2 It is clear that immune subsets of Seq generate quasi-finite linear orderings and therefore there are $2^{\aleph_0}$ quasi-finite C.O.T.s and in fact we shall show in theorem 15.2.6 that there are $2^{\aleph_0}$ quasi-finite C.O.T.s of each countably infinite order type. However, there are *recursive* quasi-finite linear orderings. This result, proved below as theorem 14.2.2, is due to S. Tennenbaum. We recall that a C.O.T. is said to be *recursive* if it contains a recursive (or equivalently r.e.) linear ordering.

14.2.1 THEOREM. If A is quasi-finite and r.e. (or recursive) then the order type of A is either finite or of the form

$$\omega + (\omega + \omega^*) \cdot \tau + \omega^*,$$

where τ is some countable order type.

PROOF. Clearly we may assume without loss of generality that $\mathsf{A} \subseteq \mathsf{R}$. Since A is r.e. and non-empty theorem 3.3.1 shows that the type of A is of the form $\omega + \sigma$ and that of A^* of the form $\omega + \tau$ where σ, τ are countable order types. Hence A is of a type $\omega + \rho + \omega^*$. Now by the proof of theorem 3.3.1, every element in C‘A has an immediate successor (predecessor) if it has any successor (predecessor). Hence the order type of A is as stated in the theorem.

14.2.2 THEOREM (R. S. TENNENBAUM). There exists a recursive linear ordering of type $\omega + \omega^*$ embeddable in R which is infinite but contains no recursive ascending or descending chain.

PROOF.[141] By corollary 1.2.6 it is sufficient to show that there exists a r.e. linear ordering with the required properties. We first sketch the main lines of the proof. Since R has type $1 + \eta$ we can identify the elements of Seq which are non-zero with rational numbers and we can identify real numbers with Dedekind sections of R. For the proof we first construct a real number A such that neither the upper nor the lower Dedekind section determined by A is r.e. We then show that this real number A is the limit of a recursive sequence of rationals (in fact it is the unique limit). Finally we show that the intersection of this sequence with the lower (upper) Dedekind section of A cannot contain a recursive ascending (descending) chain. We now proceed to the details.

Let $\omega_0, \omega_1, \ldots$ be a uniform enumeration of the r.e. sets of non-zero *sequence* numbers. (Since Seq is a recursive predicate such an enumeration is readily obtained from any of the usual enumerations of sets of natural numbers, cf. KLEENE (1952).)

We shall also use the notation Σ_i, Π_i, to express that a predicate is expressible in prenex form with i alternations of (number) quantifiers the first being existential, universal. $\Delta_i = \Sigma_i \cap \Pi_i$.

$$x \in \omega_n \quad \text{is in} \quad \Sigma_1 - \Pi_1 .$$

14.2.3 LEMMA (see KLEENE, 1952). A predicate is in Δ_2 if, and only if, it is recursive in predicates in Σ_1 (or Π_1).

14.2.4 LEMMA. The predicate $\mathrm{T}(n)$: "ω_n contains at least three distinct elements" is Σ_1.

PROOF.

$$\mathrm{T}(n) \Leftrightarrow (\exists x)\,[(x)_0 \in \omega_n \;\&\; (x)_1 \in \omega_n \;\&\; (x)_2 \in \omega_n \;\&\; (i \neq j, 0 \leq i, j \leq 2 \Rightarrow (x)_i \neq (x)_j)] .$$

For any x (be it a sequence number or not) the predicate $\mathrm{E}(y, x)$: "y extends x & $\mathrm{Seq}(y)$" defined by

$$\mathrm{E}(y, x) \Leftrightarrow \mathrm{Seq}(y) \;\&\; lh(x) \leq lh(y) \;\&\; (\forall i)\,(i < lh(x) \Rightarrow (x)_i = (y)_i)$$

is (primitive) recursive. Hence there exists a recursive function $t(n, x)$ such that $t(n, x) > n$ and $\omega_{t(n,x)} = \{y : y \in \omega_n \,\&\, \mathrm{E}(y, x)\}$.

Main construction. We define a sequence of integers a_i such that $\{\langle a_0, \ldots, a_n, 1\rangle\}_{n=0}^{\infty}$ has limit A. (We add the 1 at the end simply in order to avoid violating the definition of sequence number.)

Stage 1. $a_0 = 0$.

Stage $n+1$. Suppose $a_0, \ldots, a_{m_{n-1}}$ have been determined and $t(n, \langle a_0, \ldots, a_{m_{n-1}}\rangle) = y$ (where $y > n$).

a) If $\mathrm{T}(y)$ then enumerate ω_y until three distinct elements $u \prec v \prec w$

have been obtained. Let $m+1=\max(lh(u), lh(v), lh(w))$, ($m$ is always defined since a_0 has been determined.) If $v=\langle a_0,\ldots, a_{m_{n-1}},\ldots, a_k\rangle$ then $a_0,\ldots, a_k$ are thereby determined and if $k+1<m$ then set

$$a_{k+1}=\cdots=a_m=0 \quad \text{and} \quad m_n=m\,.$$

b) If $\neg \mathrm{T}(y)$ then set $m_n=m_{n-1}$ thereby determining no new a_i.

14.2.5 LEMMA. For each n, a_n is eventually determined and no a_n is ever changed once determined.

PROOF. a_0 is determined. If $a_0,\ldots, a_{m_{n-1}}$ have been determined then since every set is repeated infinitely many times in the enumeration ω_i we can always find 3 extensions of any given sequence number. Thus $a_0,\ldots, a_{m_{n-1}}$ has a proper extension. The first part of the lemma now follows by induction and the second assertion is trivial.

Let us define $\mathrm{A}_l=\{x:x\prec\mathrm{A}\}$ and $\mathrm{A}_u=\{x:x\succ\mathrm{A}\}$ to be the lower and upper Dedekind sections determined by A.

14.2.6 LEMMA. A_l, A_u are not r.e.

PROOF. If A_l is r.e. then $\mathrm{A}_l=\omega_n$ for some n and $\mathrm{T}(n)$ holds since A_l is infinite. Thus ω_n contains three distinct sequence numbers $u\prec v\prec w$.

Extend the sequences by 0's to u', v', w' so that $lh(u')=lh(v)$, $lh(w))=m+1$ and abbreviate m_{n-1} by r; then

$$\begin{aligned}
u' &= \langle a_0,\ldots, a_r, b_{r+1},\ldots, b_m\rangle,\\
v' &= \langle a_0,\ldots, a_r, a_{r+1},\ldots, a_m\rangle,\\
w' &= \langle a_0,\ldots, a_r, c_{r+1},\ldots, c_m\rangle, \quad \text{say}.
\end{aligned}$$

Now A lies strictly within the open interval (of real numbers) (u, v) since u', v', w' all differ and $\langle a_0,\ldots, a_n\rangle$ for $n>m$ extends v'. Hence $w\notin\mathrm{A}_l$ which is a contradiction. Similarly A_u is not r.e.

We note that since A lies in the intersection of all the intervals (u, w) the set $\{\langle a_0,\ldots, a_m, 1\rangle\}_{m=0}^{\infty}$ has precisely one limit point.

Now the procedure for constructing A is clearly recursive in Σ_1 hence by lemma 14.2.3 A is Δ_2, thus the predicate $\mathrm{A}(a)$: "A extends a" defined by

$$(\exists n)\,(n\geq lh(a)\ \&\ \mathrm{E}(\langle a_0,\ldots, a_n\rangle, a))$$

is Σ_2 and so we have

$$A(a) \Leftrightarrow (\exists x)(\forall y)\, R(a, x, y),$$

for some recursive R.

Define the set B by

$$a \in B_n \Leftrightarrow (\exists x)(\forall y)(y \leq n \Rightarrow R(a, x, y) \;\&\; lh(a) \leq n),$$

$$B = \bigcup_{A=0}^{\infty} B_n,$$

then B is clearly r.e. Since A_l, A_u are not r.e. there are infinitely many $a_i \neq 0$ hence

$$C = \{c : c \in B \;\&\; \mathrm{Seq}(c)\}$$

is an infinite r.e. set containing infinitely many sequence numbers of the form $\langle a_0, \ldots, a_m, 1\rangle$ and hence has A as a limit point. Finally, let $\mathsf{C} = [C]$ then C is the required ordering. For if C had a recursive ascending chain then C would have a limit point distinct from A. This is impossible for let L be such a limit point then L, A are different limits of the sequence $\{\langle a_0, \ldots, a_m, 1\rangle\}_{m=0}^{\infty}$.

Finally, the order type of C is $\omega + (\omega^* + \omega)\,\tau + \omega^*$ for some countable type τ by theorem 14.2.1. But if $\tau \geq 1$ then C would have a limit point distinct from A which we have shown is impossible. This completes the proof.

14.2.7 THEOREM. (Recursive) Quasi-finite C.O.T.s are closed under addition and multiplication.

PROOF. (Recursive) Quords are closed under addition and multiplication so the theorem follows from definition 14.1.1.

Summing up we get

14.2.8 THEOREM. There exist $\aleph_0$ recursive quasi-finite C.O.T.s of type $(\omega + \omega^*) \cdot n$ for every $n > 0$.

PROOF. By theorem 14.2.2 there exists a recursive quasi-finite C.O.T. A, say, of type $\omega + \omega^*$. By theorem 14.2.7 $A \cdot n$ is a recursive quasi-finite C.O.T. for every n and by corollary 2.4.12, $A \cdot n = B \cdot n \,\&\, n > 0$ implies $A = B$ so it suffices to prove there exist $\aleph_0$ recursive quasi-finite C.O.T.s of type $\omega + \omega^*$.

Let A be a r.e. quasi-finite linear ordering of type $\omega + \omega^*$ and let B be

$\mathsf{A}[\mathsf{B}$ where B is any infinite r.e. proper subset of C‘A. Since C‘A is infinite such a B exists. B is clearly a r.e. linear ordering. But if B contained a recursive ascending or descending chain then so would A hence B is r.e. and quasi-finite. But $\mathsf{B} \not\simeq \mathsf{A}$ by theorem 14.1.4. Clearly, we can then repeat the process to obtain $\aleph_0$ such B_i such that $\mathsf{A} \supset \mathsf{B}_i \supset \mathsf{B}_j$ for $i<j$, but $\mathsf{A} \not\simeq \mathsf{B}_i \not\simeq \mathsf{B}_j$ if $i \neq j$. Taking C.O.T.s yields the required result.

14.2.9 THEOREM. If A is quasi-finite then

$$A = B + A + C = 0 \Rightarrow B = C = 0 .$$

PROOF. By theorem 3.2.7, $A = B + A + C$ implies $C=0$. By theorems 3.2.7 and 2.4.2, $A = B + A$ implies $B=0$.

14.3 We defined quasi-finite C.O.T.s as those C.O.T.s A such that both A and A^* are quords, so most of our results for $\leq$ obtained in previous chapters for quords also apply to $\leq^*$ for quasi-finite C.O.T.s with the obvious exceptions caused by the failure of closure conditions, etc.

We shall omit detailed mention of the results. But we note that $\leq^*$ is reflexive, anti-symmetric and transitive on quasi-finite C.O.T.s and is a tree ordering. It is not a linear ordering or a partial well-ordering. The former follows from theorem 14.2.8 for let A be a quasi-finite C.O.T. of type $(\omega+\omega^*)\cdot 2$ then if $\mathsf{A} \in A$ there exists $\mathsf{B} \subset \mathsf{A}$ of type $\omega^*+\omega$. Clearly A and COT(B) are incomparable under $\leq^*$ since they have final segments with incomparable (classical) order types. $\leq^*$ is not a partial well-ordering for if A is a quasi-finite C.O.T. as given by theorem 14.2.8, then A is expressible in the form $n + B_n$ for each n where B_n is quasi-finite and

$$B_m <^* B_n \quad \text{if} \quad n < m ,$$

(where $A <^* B$ abbreviates $A \leq^* B \,\&\, A \neq B$) i.e. $\{B_n\}$ is a descending chain of quasi-finite C.O.T.s under $\leq^*$.

We do remark, however, that although $\subseteq$ is not anti-symmetric in general we have

14.3.1 THEOREM. $\subseteq$ is a partial ordering of the quasi-finite C.O.T.s.

PROOF. If $A \subseteq B$ and $B \subseteq A$ then there exist $\mathsf{A} \in A$, $\mathsf{B} \in B$ and recursive isomorphisms p, q such that

$$p : \mathsf{A} \simeq \mathsf{B}_1 \subseteq \mathsf{B},\; q : \mathsf{B} \simeq \mathsf{A}_1 \subseteq \mathsf{A} .$$

Hence

$$qp: \mathsf{A} \simeq \mathsf{A}_2 \subseteq \mathsf{A}.$$

By theorem 14.1.4, qp extends the identity on A and similarly pq extends the identity on B, thus $\mathsf{A} \simeq \mathsf{B}$ and $A = B$.

The other properties are obvious.

By the proof of theorem 14.2.8, $\subseteq$ is not a partial well-ordering of the quasi-finite C.O.T.'s.

CHAPTER 15

LOSOLS

15.1 We saw in the previous chapter that quasi-finite C.O.T.s do not have very reasonable properties. We now turn to losols which we consider are nice and reasonable. This chapter is devoted to examining their basic properties: some of the proofs become very simple but with more complicated proofs we obtain some more far reaching cancellation laws. However, we leave over the general problem of the theory analogous to the Myhill-Nerode theory of combinatorial functions to another occasion. Aczel has shown that the obvious analogue is not fruitful, basically because there is no obvious correspondence between numbers and order types but Nerode and the author have now found methods of circumventing these difficulties.

Since losols are special cases of C.O.T.s all the definitions given so far apply to losols and we shall not comment on any restrictions since these are obvious. We do, however, stress again that all linear orderings we consider are to be embeddable in R by means of one-one, partial recursive, order preserving maps.

15.1.1 THEOREM. If $\mathsf{A} \simeq \mathsf{B}$ and $C`\mathsf{A}$ is $\left\{\begin{matrix}\text{finite}\\ \text{immune}\end{matrix}\right.$ then $C`\mathsf{B}$ is $\left\{\begin{matrix}\text{finite}\\ \text{immune.}\end{matrix}\right.$

PROOF. Left to the reader.

15.1.2 THEOREM. The following statements are equivalent for $A = C`\mathsf{A}$.

(i) A contains an infinite r.e. subset,

(ii) A contains an infinite recursive subset,

(iii) A is not isolated.

PROOF. See theorem 13.1.4.

15.1.3 THEOREM. If C‘A is isolated then A is quasi-finite. The converse is false.

PROOF. If A is not quasi-finite then A contains a recursive (ascending or descending) chain; *a fortiori* C‘A contains an infinite r.e. subset and hence is not isolated. In theorem 14.2.2 it was shown that there exist r.e. quasi-finite linear orderings so the converse is false.

15.1.4 DEFINITION. A linear ordering A is said to be *isolated* if C‘A is isolated. A C.O.T. A is said to be a *losol* if it contains an isolated linear ordering. We write $\mathscr{L}=\{A:A \text{ is a losol}\}$.

15.1.5 COROLLARY. A is a losol$\Rightarrow A$ is quasi-finite.

PROOF. Immediate from theorem 15.1.3.

15.1.6 THEOREM. If some A$\in A$ is isolated then every A$\in A$ is isolated.

PROOF. Left to the reader.

The term "losol" was invented by Joseph Rosenstein and the author after the author had considered several other words. It has the following motivations: it suggests linearly ordered isol, which is not strictly accurate but losols are C.O.T.s of linear orderings whose fields are isolated (and embeddable in R); further, it is palindromic and losols have a lot of symmetric properties as we shall show, and finally there is a paronomasial rapport with "isol".

15.1.7 THEOREM. A recursive C.O.T. is a losol if, and only if, it is finite.

In the next section we shall give Hamilton and Nerode's proof that for each countably infinite order type τ, there exist $2^{\aleph_0}$ losols of type τ.

15.2 In this section we commence discussion on the order relations on losols. We observe that if A is a losol then A is quasi-finite so both A and A^* are quords and most of our results are applicable to $\leq^*$ too, where we now *re-define* $\leq$ and $\leq^*$ by

$$A \leq B \Leftrightarrow (\exists C)(A + C = B),$$
$$A \leq^* B \Leftrightarrow (\exists C)(C + A = B).$$

We recall that we write

$$A \subseteq B \quad \text{for} \quad (\exists \mathsf{A} \in A)(\exists \mathsf{B} \in B)(\mathsf{A} \subseteq \mathsf{B});$$

we shall also write $A \subset B$ for $A \subseteq B \,\&\, A \neq B$.

15.2.1 DEFINITION. A is said to be an initial (middle, final) segment of B if

$$A \leq B \;((\exists C)(\exists D)(C + A + D = B),\; A \leq^* B).$$

A is said to be a *weak predecessor* of B if $A \subseteq B$.

The next theorem is inspired by theorem 41 of DEKKER and MYHILL (1960).

15.2.2 THEOREM. (i) $A \leq B$ or $A \leq^* B$ implies $A \subseteq B$,

(ii) Every infinite losol has $\aleph_0$ initial (middle, finite) segments and $2^{\aleph_0}$ weak predecessors,

(iii) $A \subseteq B$ does not imply $A \leq B$ nor $A \leq^* B$.

PROOF. The first part is obvious and the third follows from the second. Since A is infinite any $\mathsf{A} \in A$ contains countably infinitely many elements in its field. Let A be fixed and for $a \in \mathrm{C}`\mathsf{A}$ let $A_a = \mathrm{COT}(a)\mathsf{A})$ then

$$A_a \leq A.$$

By part (i) of the present theorem and theorem 14.1.5 $A_a \neq A$ and $A_a \neq A_b$ if $a \neq b$. Thus there are exactly $\aleph_0$ distinct initial segments of A. The result follows for the other two types of segment since (i) initial segments are middle segments (put $C=0$) and (ii) $A - A_a$ is a final segment of A.

Since if A is a losol and for $\mathsf{A} \in A$, $\mathrm{C}`\mathsf{A}$ is infinite, $\mathrm{C}`\mathsf{A}$ has $2^{\aleph_0}$ infinite subsets. As we have argued many times before these subsets give rise to $2^{\aleph_0}$ C.O.T.s all of which must be losols.

15.2.3 THEOREM. (i)–(iii) below are equivalent and so are (i)*–(iii)*.

(i) $A \leq B$,	(i)* $A \leq^* B$,
(ii) $(\exists C)(A + C = B)$,	(ii)* $(\exists C)(C + A = B)$,
(iii) $A^* \leq^* B^*$,	(iii)* $A^* \leq B^*$.

PROOF. Left to the reader.

15.2.4 THEOREM. If A is a quord and $A \leq A^*$ or $A^* \leq A$, then $A = A^*$ (so A is quasi-finite).

PROOF. $A \leq A^*$ implies $A^* \leq^* A^{**} = A$ by theorem 15.2.3. Hence by theorem 4.2.6 (or 2.4.9) we conclude $A = A^*$.

15.2.5 THEOREM. If A is a losol and $B \subseteq A$ or $B \leq A$ or $B \leq^* A$ then B is a losol.

PROOF. Left to the reader.

15.2.6 THEOREM (HAMILTON-NERODE). Let τ be any countably infinite order type, then there exist $2^{\aleph_0}$ losols of (classical) order type τ which are pairwise incomparable under $\subseteq$.

Since C.O.T.s are equivalence classes which contain at most a countable number of linear orderings it suffices to prove the following lemma.

15.2.7 LEMMA (HAMILTON-NERODE). Let τ be a countably infinite order type, then there exists a collection $\mathscr{A} = \{A_i : i \in I\}$ of distinct subsets of Seq such that

(i) cardinal of $I = 2^{\aleph_0}$,

(ii) $\mathsf{A}_i = [A_i]$ is of order type τ,

(iii) A_i is isolated (equivalently, immune),

(iv) if $i, j \in I$, $i \neq j$, then there is no one-one partial recursive function mapping A_i into A_j.

Remark. Although the ordering $\prec$ does not apparently enter into the statement of the theorem (or lemma) in any significant way, nevertheless we have not found it possible to prove the result directly from DEKKER and MYHILL's (1960) theorem that there exist $2^{\aleph_0}$ infinite isols.

PROOF of the lemma. Let Seq′ denote Seq $- \{0\}$ and let A_τ be a subset of Seq′ with order type τ. (Such a set exists by the classical version of theorem 1.2.3.) Further, since A_τ is countable let $a_0, a_1, \ldots$ be a (possibly non-effective) enumeration of A_τ without repetitions. Let $\varphi_0, \varphi_1, \varphi_2, \ldots$ be a fixed (again non-effective) enumeration of all one-one partial recursive functions whose *ranges contain an infinite number of points of* Seq′ and let φ_0 be the (total) identity function. (We could restrict ourselves to order preserving φ to prove the theorem but this is clearly not necessary.) We now define a full binary tree whose nodes lie in Seq′. The A_i will be the branches of this tree and (i) will be satisfied automatically.

We shall make extensive use of Kleene's indefinite description operator ν which has the property that if $\mathfrak{R}$ is a partial recursive predicate then

$$(\exists x)\,\mathfrak{R}(x, y) \Rightarrow \mathfrak{R}(\nu_x \mathfrak{R}(x, y), y)$$

and $\nu_x \mathfrak{R}(x, y)$ is a partial recursive function of y uniformly in $\mathfrak{R}$.

We define our tree in stages. The nodes will be the points x_{si} where $i \in 2^s$. We shall regard 2^s as the set of all sequences of length s of zeros and ones, (well-)ordered lexicographically. Let T_{si} be the set of nodes of the tree which have been defined before x_{si} and let B_{si} be the set of nodes of the tree which have been defined before x_{si} and lie on the same branch as x_{si}, i.e. those of the form x_{rj} where $r<s$ and $i=k_0 \ldots k_s$ and $j=k_0 \ldots k_r$ ($k_i=0$ or 1). We proceed to define our construction in stages.

Stage $s\,(s \geq 0)$. For each $i \in 2^s$ define successively x_{si} to be an element t of Seq$'$ such that

(a) t stands in the same order relation (with respect to $\prec$) to the elements of B_{si} as a_s does to the elements $a_0, \ldots, a_{s-1}$,

(b) $t \in \bigcup_{0 \leq j \leq s} \delta\varphi_j$ but $(\forall j)\,(j \leq s \,\&\, t \in \delta\varphi_j \Rightarrow \varphi_j(t) \notin \mathrm{T}_{si})$,

(c) $(\forall j)\,(j \leq s \Rightarrow (\forall z)\,(z \in \mathrm{T}_{si} \cap \delta\varphi_j \Rightarrow t \neq \varphi_j(z)))$,

(d) $t \notin \{\varphi_r(z_r) : r \leq s\}$,

where

$$z_k = \nu_w \{\varphi_k(w) \text{ is defined } \& \ \varphi_k(w) \notin \{\varphi_j(z_j) : j < k\} \cup \{x_{jl} : j < k \ \& \ l \in 2^j\} \cup \{x_{kl} : l < i\},$$

where $l<i$ means l precedes i in the lexicographic ordering of 2^s.

Now writing $\overline{\overline{\mathrm{A}}}$ for the cardinal of the set A, $\overline{\overline{\mathrm{T}}}_{si}$, $\overline{\overline{\mathrm{B}}}_{si}$ are finite for every s, i and therefore, since R is dense, $\rho\varphi_0 \supseteq$ Seq$'$ and $\rho\varphi_j \cap$ Seq$'$ is infinite for all j we can always find a t satisfying (a)–(d) above. By condition (c) and the fact that φ_0 is the identity, no number chosen as t at some point can have been used previously. Now let A_i for $i \in 2^\omega$ (i.e. i an infinite sequence of zeros and ones) be

$$\{a_{si_s} : i_s = k_0 \ldots k_s \quad \text{and} \quad i = k_0 \ldots k_s \ldots,\ s = 0, 1, \ldots\}.$$

Since $\overline{\overline{2^\omega}} = 2^{\aleph_0}$ and $i \neq j$ implies $\mathrm{A}_i \neq \mathrm{A}_j$ by the remark above (and condition (c)), condition (i) of the lemma is fulfilled. Each $\mathsf{A}_i = [\mathrm{A}_i]$ has order type τ by (a) so condition (ii) is satisfied. (iii) is also satisfied since A_i contains no point of $\{\varphi_i(z_i) : i = 0, 1, \ldots\}$ and therefore A_i omits at least one point from every infinite r.e. set $\subseteq$ Seq$'$.

Suppose p is a one-one partial recursive function whose domain includes A_i and which maps A_i into A_j. Then p is φ_k for some k. If $A_i \neq A_j$ then, by the definition of the A_i as branches in the tree, condition (c) implies that for sufficiently large s, $s<t, u$ implies $x_{ti} \neq x_{uj}$ where x_{ti}, x_{uj} are the elements of A_i, A_j defined at stage t, u, respectively. Hence for some $r, t>k$ we have $\varphi_k(x_{ri})=x_{tj} \neq x_{ri}$. Since $\varphi_k(x_{ri})$ is defined, if x_{ri} is defined before x_{tj} we obtain a contradiction with condition (c) and if x_{ri} is defined after x_{tj} we obtain a contradiction with condition (b). This completes the proof.

15.3.

15.3.1 THEOREM. If A, $B \in \mathscr{L}$, then

(i) $A+B \in \mathscr{L}$,

(ii) $A \cdot B \in \mathscr{L}$, and

(iii) $A^B \in \mathscr{L}$.

PROOF. (i) follows at once from theorem 13.3.1.

(ii) Let $\mathsf{A} \in A$, $\mathsf{B} \in B$, then $C`\mathsf{A} \cdot \mathsf{B} = \{j(a, b): a \in C`\mathsf{A}, b \in C`\mathsf{B}\} = \mathsf{C}$, say. Suppose C contains an infinite r.e. set D then one of the two r.e. sets

$$A_1 = \{a: (\exists b)\,(j(a, b) \in D)\},$$

or

$$B_1 = \{b: (\exists a)\,(j(a, b) \in D)\} \text{ is infinite},$$

since otherwise C is contained in $j(X, Y)$ where X, Y are finite. It follows that $C`\mathsf{A}$ or $C`\mathsf{B}$ is not isolated.

(iii) In similar fashion to (ii) we show losols are closed under exponentiation. Let

$$\mathsf{A} \in A, \ \mathsf{B} \in B$$

and let

$$C = C`\mathsf{A}^{\mathsf{B}}.$$

By our general blanket assumption we may suppose $\mathsf{B} \subseteq \mathsf{R}$. Let us define "$b$ occurs$_1$ in n" to be short for "n is of the form

$$e\begin{pmatrix} \cdots\, b \,\cdots \\ \cdots\cdots\cdots\cdots \end{pmatrix}"$$

and similarly define "a occurs$_2$ in n" with a on the bottom line. Suppose C contains an infinite recursive subset D then let

$$A_1 = \{a: (\exists n)\,(n \in D \,\&\, a \text{ occurs}_2 \text{ in } n)\},$$

$$\mathrm{B}_1 = \{b:(\exists n)(n \in \mathrm{D} \ \& \ b \text{ occurs}_1 \text{ in } n)\},\ \mathsf{A}_1 = [\mathrm{A}_1] \text{ and } \mathsf{B}_1 = [\mathrm{B}_1].$$

Clearly A_1, B_1 are r.e. sets contained in C‘A, C‘B, respectively. We shall obtain a contradiction to the assumption that A_1, B_1 are both finite.

Suppose B_1 is finite then B_1 may be enumerated as

$$b_1 \succ b_2 \succ b_3 \cdots \succ b_k.$$

By the definition of exponentiation it follows that each bracket symbol in $(\mathsf{A}_1, \mathsf{B}_1)$ must contain $\leq k$ columns. Hence if A_1 has m elements $(\mathsf{A}_1, \mathsf{B}_1)$ contains at most

$$1 + m^k$$

bracket symbols and hence C‘$\mathsf{A}_1^{\mathsf{B}_1}$ is finite. This is the required contradiction since C‘A^{B} contains the infinite (recursive) subset D.

It is, we hope, now clear why quords are not closed under exponentiation (cf. theorem 7.2.1, (ii)). Although A, B may have no recursive descending chains, nevertheless it is possible to find (and Parikh found) an A and a B such that A^{B} contains a recursive descending chain where the length (equals number of columns) of corresponding bracket symbols *increases* as the chain descends.

15.3.2 THEOREM. If either

(i) $A + B \in \mathscr{L}$ or

(ii) $0 \neq A \cdot B \in \mathscr{L}$ or

(iii) $A^B \in \mathscr{L} \ \& \ A \neq 0, 1 \ \& \ B \neq 0$, then

$$A, B \in \mathscr{L}.$$

PROOF. If $\mathsf{A} \in A$, $\mathsf{B} \in B$ then under the hypotheses of the theorem there exist p_1, p_2, p_3 such that

$$p_1 : \mathsf{A} \subseteq \mathsf{A} + \mathsf{B},\ p_2 : \mathsf{A} \subseteq \mathsf{A} \cdot \mathsf{B},\ p_3 : \mathsf{A} \subseteq \mathsf{A}^{\mathsf{B}}$$

and similarly for B there exist q_1, q_2, q_3.

For example, let

$$p_3 : x \Rightarrow e\begin{pmatrix} 0 \\ x \end{pmatrix},$$

where we assume $0 \neq \min(\mathsf{B})$. We leave other definitions to the reader. It now follows by theorem 15.2.5. that $A, B \in \mathscr{L}$.

CHAPTER 16

ARITHMETIC LAWS FOR LOSOLS

16.1 Now we are entirely concerned with losols unless otherwise stated. In particular, with our new definition $A \leq B$ is equivalent to $(\exists C)\ (A+C= B)$. This definition does not agree with that given earlier for C.O.T.s since there are $2^{\aleph_0}$ initial segments of any ordering of type η but only $\aleph_0$ *separable* segments of such an ordering.

This chapter is concerned almost entirely with cancellation laws. The author has been inspired by DEKKER and MYHILL's monograph (1960) particularly the proof of theorem 116(a). However, it turned out that the proofs instead of being more complicated are in fact simpler than those proofs. This is mainly because of the trivial classical lemma below.

16.1.1 LEMMA. If $\mathsf{A}, \mathsf{B}, \mathsf{C}$ are finite linearly ordered sets such that C‘B, C‘C have the same number of elements and $\mathsf{A} \neq \emptyset$ then there are unique minimal order isomorphisms p, q, r with domains C‘B, C‘AB, C‘AC, respectively such that

$$p\colon \mathsf{B} \sim \mathsf{C},$$
$$q\colon \mathsf{AB} \sim \mathsf{AC},\ r\colon \mathsf{BA} \sim \mathsf{CA}$$

and

$$qj(a, b) = j(a, p(b)), \qquad rj(b, a) = j(p(b), a).$$

PROOF. Any linearly ordered, finite set is well-ordered. The existence of p, q, r now follows from theorem 8.3.4. Since q, r are order isomorphisms the proof is complete.

We leave the reader to extend this lemma to the cases of addition and exponentiation.

16.1.2 THEOREM. (i) If $A \neq 0$ and $AB = AC$, then $B = C$,

(ii) If $A \neq 0$ and $BA = CA$, then $B = C$.

PROOF. (i) Let $\mathsf{A} \in A$, $\mathsf{B} \in B$, $\mathsf{C} \in C$, $\mathrm{A} = \mathrm{C}`\mathsf{A}$, $\mathrm{B} = \mathrm{C}`\mathsf{B}$, $\mathrm{C} = \mathrm{C}`\mathsf{C}$, $\mathrm{B}' = \mathrm{C}`\mathsf{AB}$ and $\mathrm{C}' = \mathrm{C}`\mathsf{AC}$. We may assume without loss of generality that $\mathsf{A}, \mathsf{B}, \mathsf{C} \subseteq \mathsf{R}$ and $\langle 1 \rangle = 2 \in \mathrm{A}$. By hypothesis there is a recursive isomorphism p such that

$$p : \mathsf{AB} \simeq \mathsf{AC}.$$

We consider a formal system (not in standard formalization) whose formulae are of the forms

$$x \in \mathrm{A},\ x \in \mathrm{B},\ x \in \mathrm{C},\ x \in \mathrm{B}',\ x \in \mathrm{C}',$$

where x is a non-negative integer. (We shall refer to these as A-formulae, B-formulae, etc...). There is just one axiom namely $2 \in \mathrm{A}$ and there are three rules of inference:

R1. To infer $j(a, b) \in \mathrm{B}'$ from $a \in \mathrm{A}$ and $b \in \mathrm{B}$ and conversely.

R2. To infer $j(a, c) \in \mathrm{C}'$ from $a \in \mathrm{A}$ and $c \in \mathrm{C}$ and conversely.

R3. To infer $c \in \mathrm{C}'$ from $b \in \mathrm{B}'$ if $p(b)$ is defined and equals c and similarly to infer $b \in \mathrm{B}'$ from $c \in \mathrm{C}'$ if $p^{-1}(c)$ is defined and equals b.

If $\mathfrak{F}$ is such a formula we write $\mathrm{Cons}(\mathfrak{F})$ for the set of consequences of $\mathfrak{F}$ i.e. the set of formulae obtained from $\mathfrak{F}$ (and the axiom) by a finite number of applications of R1–R3. We leave the reader to verify that R1–R3 always lead from true formulae to true formulae. In particular we note that R3 can always be legitimately applied to a formula deduced only from true formulae since $\delta p \supseteq \mathrm{B}'$, $\rho p \supseteq \mathrm{C}'$.

Sometimes $\mathrm{Cons}(\mathfrak{F})$ is finite and these will give the important cases we shall use to define the recursive isomorphism from B to C. We observe first that if $\mathrm{Cons}(\mathfrak{F})$ is finite then we can decide when the whole of $\mathrm{Cons}(\mathfrak{F})$ has been generated (by applications of R1–R3) for this is so when any further application of R3 is (a) legitimate and (b) yields only a formula $b \in \mathrm{B}'$ or $c \in \mathrm{C}'$ which has already been generated. Since only a finite number of formulae have been generated at any stage this can be checked in finitely many steps.

In particular, if $\mathfrak{F}$ is true then $\mathrm{Cons}(\mathfrak{F})$ is finite since otherwise we should generate an infinite number of formulae and therefore an infinite number of A-formulae or of B-formulae or etc. This is impossible since A, B, C, B′, C′ are isolated.

If Cons($\mathfrak{F}$) is finite, let r, s, t, y, z be the number of A-, B-, C-, B′-, C′-formulae in Cons($\mathfrak{F}$). Then $rs=y$ since Cons($\mathfrak{F}$) is closed under R1, $rt=z$ since Cons($\mathfrak{F}$) is closed under R2, $y=z$ since Cons($\mathfrak{F}$) is closed under R3 and finally $s=t$ since $r\neq 0$ as Cons($\mathfrak{F}$) contains the axiom $2\in$A. Corresponding to the A-formulae we can enumerate the numbers such that $a\in$A is in Cons($\mathfrak{F}$) as

$$a_1 \prec \cdots \prec a_r \qquad \mathrm{A}^* = \{a_1, \ldots, a_r\}, \tag{1}$$

since $\prec$ is recursive. Similarly we obtain

$$b_1 \prec \cdots \prec b_s \qquad \mathrm{B}^* = \{b_1, \ldots, b_s\}, \tag{2}$$

and

$$c_1 \prec \cdots \prec c_s \qquad \mathrm{C}^* = \{c_1, \ldots, c_s\}. \tag{3}$$

Now if Cons($\mathfrak{F}$) is finite, since p is order preserving everywhere it is defined, we have

$$pj(a_i, b_j) = j(a_i, c_j) \quad \text{for} \quad 1 \leq i \leq r, 1 \leq j \leq s \tag{4}$$

(s may be zero, though $r \geq 1$, but this makes no difference to the argument). If Cons($\mathfrak{F}$) is finite, then p is an order isomorphism between finite sets and hence, by the lemma, unique.

Finally, we define

$$q(x) = lpj(2, x)$$

if Cons($x\in$B) is finite, undefined otherwise.

Now, if Cons($x\in$B) is finite, Cons($x\in$B) consists of $x\in$B, $2\in$A, $c\in$C, $j(2,x)\in B'$ and $j(2, c)\in$C′ for some (unique) c, hence q is one-one. It is clearly order-preserving by (4). q maps B onto C for if $y\in$C then $lp^{-1}j(2,y)$ is defined and equals x, say. But then Cons($x\in$B) is finite and $q(x)=y$. This completes the proof of (i). (ii) follows at once by permuting AB, AC where necessary with the appropriate corresponding notational changes.

By theorem 14.1.4 we know that if A, B, C are linearly ordered isolated sets then there is a minimal p such that p': AB $\simeq$ AC implies p' extends p so, on C‘B, the q we obtained is uniquely determined. However, in determining (a suitable) q we only use finite sub-functions of p. We shall use this fact to establish other cancellation laws. We observe that if A, B do not have isolated fields then there may be many recursive isomorphisms from LA onto LB even if L is finite. For, classically, we know that $\eta+1+\eta=\eta$ and $2\cdot\eta=2\cdot(\eta+1+\eta)=\eta$ and we can map any recursively linearly ordered set of type η onto itself in $\aleph_0$ distinct ways.

16.2.

16.2.1 THEOREM. (i) If $A \neq 0$ and $AB \subseteq AC$ then $B \subseteq C$.

(ii) If $A \neq 0$ and $BA \subseteq CA$ then $B \subseteq C$.

PROOF. Exactly as for theorem 16.1.2 except we omit the part showing that the q constructed is onto.

16.2.2 THEOREM. (i) If $A \neq 0$ and $AB \leq AC$, then $B \leq C$,

(ii) If $A \neq 0$ and $BA \leq CA$, then $B \leq C$.

PROOF. Let A, B, C be as in the proof of theorem 16.1.2. Construct q as in that proof then it is sufficient to show that (a) q maps B onto an initial segment of C and (b) q(B) is separable from its complement $\text{C}'' = \text{C}`\text{C} - \text{C}`q(\text{B})$.

(a) Clearly it is sufficient to show that if $c \in \text{C}$ and there exist $c' \in \text{C}$ and $b' \in \text{B}$ such that

$$c <_{\text{C}} c' \quad \text{and} \quad q(b') = c',$$

then there exists $b \in \text{B}$ with $q(b) = c$.

Let c', b' be as above. Then

$$j(2, c) <_{\text{AC}} j(2, c') = pj(2, b'),$$

so $j(2, c) = pj(a, b)$ for some $a \in \text{A}$, some $b \in \text{B}$ with $b \prec b'$. Now from the proof of theorem 16.1.2 we have

$$pj(a_1, b_1) = j(a_1', c_1)$$

implies $a_1' = a_1$ so

$$j(2, c) = pj(2, b)$$

and it now follows that

$$q(b) = lpj(2, b) = c,$$

as required.

(b) By our hypothesis $AB \leq AC$ it follows that if we set $\text{AC}[(j(\text{A}, \text{C}) - j(\text{A}, \text{B})) = \text{D}$ then AB) (D. But $x \in \text{B} \Leftrightarrow j(2, x) \in j(\text{A}, \text{B})$ and $x \in \text{C}'' \Leftrightarrow j(2, x) \in j(\text{A}, \text{C}`\text{D})$. (*)

Finally, $x \in \mathrm{B} \Leftrightarrow q(x) \in \mathrm{C}$ so we obtain

$$x \in q(\mathrm{B}) \Leftrightarrow j(2, q^{-1}(x)) \in j(\mathrm{A}, \mathrm{B}),$$

which with (*) gives $q(\mathrm{B}))(\mathrm{C}''$. This completes the proof of (i) and the proof of (ii) follows at once by permuting AB, AC etc. as in the proof of theorem 16.1.2, (ii).

16.2.3 THEOREM. (i) If $A \neq 0$ and $AB \leq^* AC$, then $B \leq^* C$.
(ii) If $A \neq 0$ and $BA \leq^* CA$ then $B \leq^* C$.

PROOF. Exactly as for theorem 16.2.2. reading "<" for ">" in the ordering relations.

16.2.4 COROLLARY. (i) $AB < AC \Rightarrow B < C$,
(ii) $BA < CA \Rightarrow B < C$,
(iii) $AB <^* AC \Rightarrow B <^* C$,
(iv) $BA <^* CA \Rightarrow B <^* C$.

PROOF. Immediate from the fact that $A < B \Leftrightarrow A \leq B \,\&\, A \neq B$ and similarly for $<^*$.

These results are in marked contrast to those for isols. The principal reason for this is that here we have our classical lemma 16.1.1 whereas its cardinal analogue is false: there are lots of isomorphisms in general in the cardinal case.

16.3.

16.3.1 THEOREM. $B \leq C$ implies neither $B + A \leq C + A$ nor $BA \leq CA$ nor $(A \neq 0 \Rightarrow B^A \leq C^A)$.

PROOF. Let $A = \mathrm{COT}(\mathsf{A})$ where A is an immune set with order type ω (such an A exists by theorem 15.2.6 or by taking the natural ordering by magnitude of integers on an immune set). Since $A \neq W$ and $n + A \leq A \Leftrightarrow W \leq A$ it follows that $m + A \nleq n + A$ if $m < n$. Taking $B = 1$, $\mathrm{C} = 2$, $|BA| = \omega = |CA|$ but $BA \leq CA$ then implies, by corollary 5.3.5 that $BA = CA$, i.e. $2A = A$. It now follows from lemma 7.4.1 and the fact that $2^W = W$ that W divides A which is impossible since A is a losol.

Finally, let $B = 2$, $C = A$ where A is as before. Then $B^A = 2^A$ so $|B^A| = \omega$ and $C^A = A^A$. Hence A^A has an initial segment of type ω which is therefore

A. We conclude that if $B^A \leq C^A$ then $A = 2^A$ whence by lemma 11.2.3 we obtain $A = W$ which is false.

The converse implications have, of course, been established earlier since all losols are quords. We also established laws like $A^B \leq A^C$ for $B \leq C$ and $A \geq 1$.

16.3.2 THEOREM. (i) $A^B = A^C$ and $A \supseteq 2$ imply $B = C$,
(ii) $A \neq 0 \,\&\, (1+B)^A = (1+C)^A$ imply $B = C$,
(iii) $A^B < A^C$ implies $B < C$,
(iv) $B^A < C^A$ implies $B < C$ if $A \neq 0$
and similarly with "$<^*$", "$\subseteq$" replacing "$<$".

PROOF. The proofs are slight modifications of the proof of the theorems in sections 16.1, 2. We assume A, B, C etc. defined as before, $\mathrm{B}' = \mathrm{C}`\mathrm{A}^{\mathrm{B}}$, $\mathrm{C}' = \mathrm{C}`\mathrm{A}^{\mathrm{C}}$ and we also assume $2 \in \mathrm{A}$, $4 \in \mathrm{A}$, $2 \in \mathrm{B}$ and $2 \in \mathrm{C}$. By the definition of exponentiation $0 \in \mathrm{B}'$, $0 \in \mathrm{C}'$ but we may assume $4 \neq \min(\mathrm{A})$.

Now we define Cons($\mathfrak{F}$) by using the (new) rules R1–R3 below.
Let $\mathrm{A}' = \mathrm{A} - \{\min(\mathrm{A})\}$
R1. To infer

$$e\begin{pmatrix} b_0 \dots b_n \\ a_0 \dots a_n \end{pmatrix} \in \mathrm{B}'$$

from $a_0 \in \mathrm{A}', \dots, a_n \in \mathrm{A}'$, $b_0 \in \mathrm{B}, \dots, b_n \in \mathrm{B}$ and conversely.

R2. To infer

$$e\begin{pmatrix} c_0 \dots c_n \\ a_0 \dots a_n \end{pmatrix} \in \mathrm{C}'$$

from $a_0 \in \mathrm{A}', \dots, a_n \in \mathrm{A}'$, $c_0 \in \mathrm{C}, \dots, c_n \in \mathrm{C}$ and conversely.

R3. To infer $c \in \mathrm{C}'$ from $b \in \mathrm{B}'$ if $p(b)$ is defined and equals c and similarly to infer $b \in \mathrm{B}'$ from $c \in \mathrm{C}'$ if $p^{-1}(c)$ is defined and equals b.

We take $2 \in \mathrm{A}$, $4 \in \mathrm{A}$ as axioms. Cons($\mathfrak{F}$) is now defined to be the set of consequences of $\mathfrak{F}$ under our new rules (and axioms). If $\mathfrak{F}$ is true then, as before, Cons($\mathfrak{F}$) contains only true formulae. Again, as previously, Cons($\mathfrak{F}$) is finite if $\mathfrak{F}$ is true and we can decide when Cons($\mathfrak{F}$) has been completely enumerated.

If Cons($\mathfrak{F}$) is finite, let r, s, t, y, z be defined as before. Then $r^s = y+1$, $r^t = z+1$, $y = z$ and hence, since $r > 1$, $s = t$. (The reason we have to take $y+1$, $z+1$ rather than y, z is that the minimum element of E^{F}

when E, F are finite exists and is 0 and we do not have $0 \in B'$, $0 \in C'$ in Cons$(\mathfrak{F})$).

As before we obtain (1), (2), (3) and since p is order preserving everywhere it is defined, we also have

$$pe\begin{pmatrix} b_{i_1} \dots b_{i_u} \\ a_{j_1} \dots a_{j_u} \end{pmatrix} = e\begin{pmatrix} c_{i_1} \dots c_{i_u} \\ a_{j_1} \dots a_{j_u} \end{pmatrix} \tag{4}$$

and

$$b_{i_1} >_{\mathrm{B}} \dots >_{\mathrm{B}} b_{i_u} \quad \text{and} \quad c_{i_1} >_{\mathrm{C}} c \dots >_{\mathrm{C}} c_{i_u}$$

for all possible bracket symbols obtainable from A*, B*, C*.

If Cons$(\mathfrak{F})$ is finite then p is an order isomorphism between finite sets (here we do not need to bother about adding 0 as minimum element) and hence, by lemma 16.1.1, unique.

Finally we define

$$q(x) = k\left(\left(pe\begin{pmatrix} x \\ 4 \end{pmatrix}\right)_0\right)$$

if Cons$(x \in \mathrm{B})$ is finite, undefined otherwise. The rest of the proof now follows analogously to that of theorems 16.1.2 *et seq.* and we are through.

16.4 This section is devoted to the proof of a cancellation metatheorem for losols. The idea of the proof is just like the ones for previous cancellation theorems, namely, given $\mathsf{P}(\mathsf{X}) \subseteq \mathsf{P}(\mathsf{Y})$ we look at all possible elements (effectively) "generated" from a single element $x \in \mathrm{C}`\mathsf{X}$ and then, since $\mathrm{C}`\mathsf{X}$ is isolated this set is finite so we can use the corresponding cancellation theorem for finite sets (or numbers) and thence obtain a recursive isomorphism from X into Y.

Notation: $\mathrm{U}_i^n(x_1, \dots, x_n) = x_i$.

16.4.1 LEMMA. If p is a (number-theoretic) non-constant one-place function obtained by finite compositions of the functions

$$\mathrm{U}_i^n,\ x + y,\ x \cdot y,\ (1 + x)^y - 1$$

and (finite) parameters, then

$$p(x) \leq p(y) \quad \text{if, and only if,} \quad x \leq y.$$

PROOF. We proceed by induction on the complexity of the function p.

Clearly U_i^n has the required property when all arguments except the ith are held constant. All the other functions listed are strictly monotone increasing in each argument or constant so they satisfy the condition. Now suppose

$$h(x) = \mathrm{U}_i^n(f_1(x, \bar{a}_1), \ldots, f_n(x, \bar{a}_n)),$$

where the f_i are functions obtained by composition from the given functions and the $\bar{a}_i$ are sequences $a_{i1}, \ldots, a_{ik_i}$ of finite parameters. Then $h(x) = f_i(x, \bar{a}_i)$ so this case reduces to a simpler one.

Finally suppose

$$h(x) = g(f_1(x, \bar{a}_1), f_2(x, \bar{a}_2)),$$

where at most one of f_1, f_2 is constant and the rest of f_1, f_2 satisfy the condition. Then $h(x)$ is strictly monotonic increasing since it is increasing in both arguments and the f_i are increasing too. Hence

$$x \leq y \Leftrightarrow h(x) \leq h(y).$$

We note that these are the only cases we have to consider since we do not need to distinguish between constant functions and parameters as far as their *values* are concerned.

16.4.2 THEOREM (CANCELLATION METATHEOREM FOR LOSOLS). If P is a non-constant one-place function (from losols to losols) obtained by finite compositions of the functions

$$U_i^n,\ X + Y,\ X \cdot Y,\ (1 + X)^Y - 1,\ X^*$$

and (losol) parameters, then

(i) $P(X) \subseteq P(Y)$ if, and only if, $X \subseteq Y$ and

(ii) $P(X) = P(Y)$ if, and only if, $X = Y$.

PROOF. Suppose the losol parameters involved in P are $A_1, \ldots, A_n$. Then, since P is non-constant we must have certain of the $A_i > 0$. We shall assume that in this case if $\mathsf{A}_i \in A_i > 0$ then $\min(\mathsf{A}_i) = 0$. We shall also assume that all the linear orderings we consider are $\subseteq \mathsf{R}$ but that $\min(\mathsf{X})$, $\min(\mathsf{Y}) \neq 0$. We associate with P a set of *canonical embeddings* of X in $\mathsf{P}(\mathsf{X})$. P is a function from linear orderings to linear orderings inducing P. We shall assume without loss of generality that P is obtained by composition analogously to P where the basic functions correspond as below.

U_i^n	U_i^n,	
$X+Y$	$\mathsf{X}+\mathsf{Y}$	(see theorem 2.2.2),
$X\cdot Y$	$m(\mathsf{X}, \mathsf{Y})$	(see theorem 6.1.6),
$(1+X)^Y-1$	$e(\mathsf{X}, \mathsf{Y})$	(see below),
X^*	$r(\mathsf{X})$	(see below).

Let R' denote $\mathsf{R}[(\mathrm{Seq}-\{0\})$ then R', R'^* are recursive dense linear orderings without first or last elements and so by theorem 1.2.4, for some r, $r{:}\,\mathsf{R}'^*\simeq\mathsf{R}'$. We take r to be fixed. This is the r of the table above.

Since we are assuming $\min(\mathsf{X})\neq 0$ we can take as representative of $1+X$, $\mathsf{X}^+=\{\langle 0,0\rangle\}\hat{+}\mathsf{X}$.
Let $e'(\mathsf{X}, \mathsf{Y})$ be the ordering $(\mathsf{X}^+)^{\mathsf{Y}}$ without its first element $\min(\mathsf{X}^+)^{\mathsf{Y}}=0$. Then $f{:}\,e'(\mathsf{R}, \mathsf{R}')\simeq\mathsf{R}'$ for some f by theorem 1.2.4 since $e'(\mathsf{R}, \mathsf{R}')$ is recursive, dense without first or last element. Now we set $e(\mathsf{X}, \mathsf{Y})=fe'(\mathsf{X}, \mathsf{Y})$. e is fixed and this is the e of the table above.

Let us write $\mathbf{1}=\{\langle 0,0\rangle\}$ and abbreviate $\mathsf{A}_1,\ldots,\mathsf{A}_n$ by $\bar{\mathsf{A}}$. We shall also, as we have often done before, write A for C‘A, etc. without comment; but we shall extend this notation in the obvious way to write P(X) for C‘$\mathsf{P}(\mathsf{X})$, etc.

Now we define the sets of canonical embeddings $\mathrm{C}(F)$ associated with F as follows.

$$\mathrm{C}(U_i^n)=\{\mathrm{U}_i^n\}\quad (\mathrm{U}_i^n(x_1,\ldots,x_n)=x_i),$$
$$\mathrm{C}(F_1(X)+F_2(X))=\{\hat{0}p_1(x), \hat{1}p_2(x){:}\, p_1\in\mathrm{C}(F_1), p_2\in\mathrm{C}(F_2)\},$$
$$\mathrm{C}(F_1(X)\cdot F_2(X))=\{m(p_1(x), p_2(x)){:}\, p_1\in\mathrm{C}(F_1)\ \&\ p_2\in\mathrm{C}(F_2)\},$$
$$\mathrm{C}((1+F_1(X))^{F_2(X)}-1)=\{e(p_1(x), p_2(x)){:}\, p_1\in\mathrm{C}(F_1)\ \&\ p_2\in\mathrm{C}(F_2)\},$$
$$\mathrm{C}(F(X)^*)=\{rp(x){:}\, p\in\mathrm{C}(F)\},$$
$$\mathrm{C}(A_i)=\emptyset,$$
$$\mathrm{C}(F(X)+A)=\{\hat{0}p(x){:}\, p\in\mathrm{C}(F)\},$$
$$\mathrm{C}(A+F(X))=\{\hat{1}p(x){:}\, p\in\mathrm{C}(F)\},$$
$$\mathrm{C}(F(X)\cdot A)=\{m(p(x), a){:}\, p\in\mathrm{C}(F)\},$$
$$\mathrm{C}(A\cdot F(X))=\{m(a, p(x)){:}\, p\in\mathrm{C}(F)\},$$
$$\mathrm{C}((1+F(X))^A-1)=\{e(p(x), a){:}\, p\in\mathrm{C}(F)\},$$
$$\mathrm{C}((1+A)^{F(X)}-1)=\{e(a, p(x)\}{:}\, p\in\mathrm{C}(F)\}.$$

Notes. 1. Since P is non-constant, in the cases of multiplication and exponentiation we take a to be a fixed element of a fixed $\mathsf{A}\in A$ and observe that such an a exists whenever that function is essentially involved in the definition of P since we saw above that the appropriate A_i were non-zero losols.

2. Since it is clear that parameters can be put in one at a time in any order it follows that $C(P)$ is defined uniquely, independent of the order of the parameters A_i.

3. If $p\in C(P)$ then $0\notin \mathrm{X}$ implies $0\notin p(\mathrm{X})$.

We leave the reader to prove that $p:\mathsf{X}\simeq p(\mathsf{X})\subseteq \mathsf{P}(\mathsf{X})$ and p, p^{-1} are one-one and order preserving wherever they are defined.

It is also trivial that if $\mathsf{X}\subseteq\mathsf{Y}$ then $p(\mathsf{X})\subseteq p(\mathsf{Y})$.

The implication (in the theorem) $X\subseteq Y$ implies $P(X)\subseteq P(Y)$ now follows easily by considering all the embeddings where the constants a in the definition of $C(F)$ above are allowed to range over the whole of the (appropriate) A_i. We leave this trivial verification to the reader.

Now suppose $P(X)\subseteq P(Y)$ where P is as above and the A_i are the fixed representatives of the losols A_i. By hypothesis there is a one-one partial recursive function f which is order preserving wherever it is defined and such that

$$f:\mathsf{P}(\mathsf{X})\simeq \mathsf{Z}\subseteq \mathsf{P}(\mathsf{Y})\subseteq \mathsf{R},$$

where

$$\mathsf{P}(\mathsf{X}),\ \mathsf{X},\ \mathsf{Y}\subseteq \mathsf{R}.$$

We make one further assumption on the A_i, namely, that we can choose $a=2=\langle 1\rangle\ddagger 0$ in the construction of $C(P)$. This is merely to avoid excessive complication in the notation and clearly causes no loss of generality.

Now given any *finite* subsets $\mathrm{X}'\subseteq\mathrm{X}$, $\mathrm{A}_i'\subseteq\mathrm{A}_i$, $\mathsf{P}(\mathsf{X}',\overline{\mathsf{A}}')$ is defined and finite where $\mathsf{A}_i'=[\mathrm{A}_i']$, the linear ordering generated by $\mathrm{A}_i'\subseteq\mathrm{Seq}$. This is because all the given functions lead from finite sets to finite sets. Moreover, if $y\in C`\mathsf{P}(\mathsf{Y},\overline{\mathsf{A}})$ then $y\in C`\mathsf{P}(\mathsf{Y}',\overline{\mathsf{A}}')$ for some *minimum finite* Y', A_i' $(i=1,\ldots,n)$ which are effectively computable from the description of P and y. We leave the reader to verify this by induction.

For example if

$$y\in C`(\mathsf{1}\ddagger\mathsf{A})^{\mathsf{Y}},$$

then (uniquely)

$$y=e\begin{pmatrix} y_1 \ldots y_m \\ a_1 \ldots a_m\end{pmatrix},$$

where $y_1 \succ y_2 \succ \cdots \succ y_m$ and the $a_i \in A$; so let $\mathsf{Y}' = [\{y_1, \ldots, y_m\}]$, $\mathsf{A}' = [\{a_1, \ldots, a_m\}]$. We observe that this process is uniform since $\min(\mathbf{1} \mathbin{\hat{+}} \mathsf{A})$ is independent of A.

We write

$$\mathsf{Y}' = \mathsf{P}^e_{\mathrm{Y}}(y), \quad \mathsf{A}'_i = \mathsf{P}^e_{\mathrm{A}_i}(y) \quad \text{and} \quad \mathrm{P}^e_{\mathrm{Y}}(y) = \mathrm{C}`\mathsf{P}^e_{\mathrm{Y}}(y), \quad \text{etc.}$$

Now if a necessary and sufficient condition that $P(X, \bar{A})$ be non-constant is that $A_{i_1}, \ldots, A_{i_m} > 0$ and $A_{j_1}, \ldots, A_{j_r} > 1$ then we assume $\min(\mathsf{A}_{i_k}) = 0, k = 1, \ldots, m$; $\min(\mathsf{A}_{jk}[\mathrm{A}_{jk} - \{0\}]) = \langle 1 \rangle$ $(=2)$. Clearly this can be done without loss of generality.

Now we set up our formal system. We have X-, Y-, P(X)-, P(Y)- and A_i-formulae $(i = 1, \ldots, n)$. The axioms are all formulae of the form

$$0 \in \mathrm{A}_i, \quad 2 \in \mathrm{A}_j$$

which are true. So there is a finite number of axioms.

For brevity we write $x_1, \ldots, x_m \in \mathrm{X}$ or even $\bar{x} \in \mathrm{X}$ instead of

$$x_1 \in \mathrm{X}, \ldots, x_m \in \mathrm{X}.$$

The rules are as follows:

R1a. To infer $x \in \mathrm{P}(\mathrm{X})$ from $x_1, \ldots, x_m \in \mathrm{X}, a_{11}, \ldots, a_{1j_1} \in \mathrm{A}_1; \ldots; a_{n1}, \ldots, a_{nj_n} \in \mathrm{A}_n$ if $x \in \mathrm{P}(\mathrm{X}', \bar{\mathrm{A}}')$ where

$$\mathrm{X}' = \{x_1, \ldots, x_m\}, \mathrm{A}'_i = \{a_{i1}, \ldots, a_{ij_i}\} \quad (i = 1, \ldots, n).$$

R1b. To infer $x_1, \ldots, x_m \in \mathrm{X}; a_{11}, \ldots, a_{1j_1} \in \mathrm{A}_1; \ldots; a_{n1}, \ldots, a_{nj_n} \in \mathrm{A}_n$ from $x \in \mathrm{P}(\mathrm{X})$ if $\{x_1, \ldots, x_m\} \subseteq \mathrm{P}^e_{\mathrm{X}}(x)$, $\{a_{i1}, \ldots, a_{ij_i}\} \subseteq \mathrm{P}^e_{\mathrm{A}_i}(x)$ $(i = 1, \ldots, n)$.

R2. As for R1 (a and b) with x, X replaced by y, Y.

R3. To infer $y \in \mathrm{P}(\mathrm{Y})$ from $x \in \mathrm{P}(\mathrm{X})$ if $p(x)$ is defined and equals y and similarly to infer $x \in \mathrm{P}(\mathrm{X})$ from $y \in \mathrm{P}(\mathrm{Y})$ if $p^{-1}(y)$ is defined and equals x.

As usual Cons$(\mathfrak{F})$ is the set of consequences of $\mathfrak{F}$ and the axioms (if any) under R1–R3. As in previous proofs of this type the rules preserve truth and if $\mathfrak{F}$ is true then because all the A_i, X and Y are isolated we can generate Cons$(\mathfrak{F})$ and decide (recursively) when the whole of Cons$(\mathfrak{F})$ has been generated. If Cons$(\mathfrak{F})$ is finite then let the numbers such that $a \in \mathrm{A}_i$ is in Cons$(\mathfrak{F})$ be enumerated as

$$a_{i1} \prec \cdots \prec a_{ij_i} \qquad \mathrm{A}'_i = \{a_{i1}, \ldots, a_{ij_i}\} \quad i = 1, \ldots, n.$$

Similarly we have enumerations

$$x_1 \prec \cdots \prec x_r \qquad \mathrm{X}' = \{x_1, \ldots, x_r\} \quad \text{for X-formulae},$$

$$y_1 \prec \cdots \prec y_s \qquad Y' = \{y_1, \ldots, y_s\} \quad \text{for Y-formulae},$$
$$p_1 \prec \cdots \prec p_u \qquad P' = \{p_1, \ldots, p_u\} \quad \text{for P(X)-formulae},$$
$$q_1 \prec \cdots \prec q_v \qquad Q' = \{q_1, \ldots, q_v\} \quad \text{for P(Y)-formulae}.$$

Now by R1, R2, respectively,

$$\mathsf{P}' = \mathsf{P}(\mathsf{X}', \bar{\mathsf{A}}'), \quad \mathsf{Q}' = \mathsf{P}(\mathsf{Y}', \bar{\mathsf{A}}')$$

and by R3
$f:\mathsf{P}' \simeq \mathsf{Q}'$ (or more precisely, a finite restriction of $f:\mathsf{P}' \simeq \mathsf{Q}'$). It follows at once that $u=v$. Now let $c(r, \bar{j})$ be the cardinality of $P(X, \bar{A})$ where X, A_i have cardinalities r, j_i, respectively. Then c is a number-theoretic function which satisfies the hypotheses of lemma 16.4.1 (proof by induction on the complexity of P) since $0 \in A_i$, $2 \in A_j$ are always in Cons$(\mathfrak{F})$ whenever they are required to ensure P is non-constant. Hence, if s is the cardinality of Q', $c(r, \bar{j}) = u = v = c(s, \bar{j})$ implies $x=y$ by the lemma and hence $r = s$.

We now show that if p is any canonical embedding (though we need only consider a particular one) then $f:p(\mathsf{X}') \simeq p(\mathsf{Y}')$. We then define $g(x) = p^{-1}fp(x)$ from which it follows at once that

$$g:\mathsf{X} \simeq \mathsf{Y}.$$

Since (as we have just showed) $f:\mathsf{P}(\mathsf{X}') \simeq \mathsf{P}(\mathsf{Y}')$ and $\mathsf{P}(\mathsf{X}')$, $\mathsf{P}(\mathsf{Y}')$ are finite it suffices to prove the above assertion for finite X, Y. But this is clear since given any canonical embedding p then we can explicitly compute from the description of p and the number of elements in X and the A_i which element (in the ordering $\prec$ of $\mathsf{P}(\mathsf{X})$) $p(x)$ is, where $x \in X$, say, the n_xth. Since we only used the *number* of elements of X in this calculation (and the numbers of elements in X and the A_i are computable) it follows that if x, y are the mth elements of X, Y, respectively, then $p(x), p(y)$ are the n_xth elements of $\mathsf{P}(\mathsf{X})$, $\mathsf{P}(\mathsf{Y})$, respectively. But $f:\mathsf{P}(\mathsf{X}) \simeq \mathsf{P}(\mathsf{Y})$ is an order-preserving map between finite sets so $fp(x) = p(y)$. This completes the proof.

We shall develop this technique further in our forthcoming work with Nerode.

APPENDIX A

A MIXED CANCELLATION THEOREM

by P. H. G. ACZEL and J. N. CROSSLEY

A.1 DEFINITION. If $F: \mathscr{R} \to \mathscr{R}$, F is one-one and

$$A \leq B \Leftrightarrow F(A) \leq F(B),$$

then F is said to be an *order embedding* where $\leq$ is used in the *strong* sense that $A \leq B \Leftrightarrow (\exists C)\ (A+C=B)$.[A1]

Let $S_C(A) = C+A$, $P_C(A) = C \cdot A$. By corollary 3.2.9, if C is a quord then S_C is an order embedding.

In this appendix we show that if C is a quord such that $C \geq 1$ then P_C is also an order embedding. The proof of theorem A.4 below was inspired by SIERPINSKI (1948).

A.2 LEMMA. If $f: \mathsf{C} \simeq \mathsf{X} \subseteq \mathsf{C}'$ where C' is a proper initial segment of C then C is *not* a quasi-well-ordering.

PROOF. The lemma is merely a restatement of lemma 3.2.6.

A.3 LEMMA. If $A+B=A'+B'$ is a quord, then $B<B'$ implies $A'<A$.

PROOF. Assume the hypothesis. By the directed refinement theorem (2.3.2) either

$$A \leq A' \quad \text{and} \quad B' \leq^* B,$$

or

$$A' \leq A \quad \text{and} \quad B \leq^* B'.$$

If $B<B'$, then $B' \nleq^* B$ by theorem 2.4.9. Hence $A' \leq A$ and $B \leq^* B'$. But if $A'=A$, then by corollary 3.2.9, $B=B'$ so we conclude $A'<A$.

A.4 THEOREM. If $C \geq 1$ is a quord and $C_1, C_2 < C$, then

$$CA + C_1 = CB + C_2 \quad \text{implies} \quad A = B.$$

PROOF. Let $\mathsf{A} \in A$, $\mathsf{B} \in B$, $\mathsf{C} \in C$ and let C_i $(i=1, 2)$ be initial segments of C. We may assume $\mathsf{A}, \mathsf{B}, \mathsf{C} \subseteq \mathsf{R}$ and $\min(\mathsf{C})=0$. As usual we set $\mathrm{A}=\mathsf{C}$'A, etc. Let $\mathsf{A}'=\mathsf{A}+\{\langle 0, 0\rangle\}$, $\mathsf{B}'=\mathsf{B}+\{\langle 0, 0\rangle\}$. By hypothesis there is a recursive isomorphism

$$p' : m(\mathsf{C} \cdot \mathsf{A}) + m(\mathsf{C}_1 \cdot \{\langle 0, 0\rangle\}) \simeq m(\mathsf{C} \cdot \mathsf{B}) + m(\mathsf{C}_2 \cdot \{\langle 0, 0\rangle\})$$

(cf. the proof of theorem 6.1.6). In order to simplify the notation however we shall consider instead a recursive isomorphism

$$p : \mathsf{C} \cdot \mathsf{A} \mp \mathsf{C}_1 \cdot \{\langle 0, 0\rangle\} \simeq \mathsf{C} \cdot \mathsf{B} \mp \mathsf{C}_2 \cdot \{\langle 0, 0\rangle\}$$

and leave the reader to observe that from our construction of the function $q : \mathsf{A} \sim \mathsf{B}$ it readily follows that we can construct $q' : \mathsf{A} \simeq \mathsf{B}$ such that q' is order preserving wherever it is defined. This is because the natures of A, B do not enter into the arguments insofar as their recursive properties are concerned.

Let

$$q_1(a) = lpj(0, a), \quad q_2(b) = lp^{-1}j(0, b).$$

We shall eventually define $q : \mathrm{A} \to \mathrm{B}$ such that either q is a sub-function of q_1 or q^{-1} is a subfunction of q_2.

The statement of the theorem is clearly symmetric so we consider the pairs

$$\mathsf{A}, \mathsf{B};\ \mathsf{C}_1, \mathsf{C}_2;\ p, p^{-1};\ \mathsf{C}, \mathsf{C};\ q_1, q_2;\ \text{etc.}$$

as *dual* (not in the sense of § 2.4). The dual of a statement is obtained by replacing any term by its dual. We shall make frequent use of the fact that each lemma or definition has a dual.

By lemma A.5 below it will be clear that q_2 is one-one on B and maps B into A preserving order. Dually, q_1 is one-one on A and maps A into B preserving order. The major part of the proof is devoted to showing that q_1 maps A onto B and hence (dually) that q_2 maps B onto A. The remainder of the proof shows that either q_1 or q_2 has a one-one partial recursive sub-function defined on A or B respectively.

The following notions are useful. We say $b \in \mathrm{B}$ is *generated* if $b = q_1(a)$ for some $a \in \mathrm{A}'$. Dually we define the generated elements of A.

Define the functions $f_{a,b}$ by

$$f_{a,b}(c) = c' \Leftrightarrow pj(c', a) = j(c, b).$$

It should be clear that these functions have the following properties:
1) $f_{a,b}$ is one-one, partial recursive,
2) $a \in \mathrm{A}$, $b \in \mathrm{B}$ imply $f_{a,b}$ is order-preserving (in particular on C),
3) $a \in \mathrm{A}'$, $b=0$ imply $f_{a,b}$ is order-preserving (in particular on C_2),
4) $a=0$, $b \in \mathrm{B}'$ imply $f_{a,b}^{-1}$ is order-preserving (in particular on C_1).
We shall make use of these facts without referring to them.

A.5 Lemma. Let $a \in \mathrm{A}'$, $c' <_{\mathsf{C}} c$ such that if $a=0$ then $c \in \mathrm{C}_1$. Then

$$pj(c, a) = j(c_0, b_0) \text{ and}$$
$$pj(c', a) = j(0, b_1)$$

imply $b_0 = b_1$ and $0 <_{\mathsf{C}} c_0$.

Proof. Assume the hypothesis. Then $j(0, b_1) <_{\mathsf{CB}'} j(c_0, b_0)$, so $b_1 \leq_{\mathrm{B}'} b_0$. Suppose $b_1 <_{\mathrm{B}'} b_0$, then

$$f_{a, b_1} : \mathsf{C} \simeq \mathsf{X} \subseteq \mathsf{C}[c \quad \text{for some } \mathsf{X},$$

which by lemma A.2 contradicts the fact that C is quasi-well-ordered. Hence $b_1 = b_0$ and then trivially we must have $0 <_{\mathsf{C}} c_0$ (since $0 = \min(\mathsf{C})$).

If X is a linear ordering set $x \dashv_{\mathsf{X}} y$ if, and only if, $x <_{\mathsf{X}} y$ and there is no z such that $x <_{\mathsf{X}} z <_{\mathsf{X}} y$ i.e. if, and only if, y is the immediate successor of x in X.

A.6 Lemma. Let $a \in \mathrm{A}'$, $0 <_{\mathsf{C}} c$ such that if $a=0$ then $c \in \mathrm{C}_1$. Then

$$pj(c, a) = j(0, b_0) \text{ and}$$
$$pj(0, a) = j(c_1, b_1)$$

imply $0 <_{\mathsf{C}} c_1$ and $b_1 \dashv_{\mathrm{B}'} b_0$.

Proof. Clearly $c_1 \in \mathrm{C}$ and $b_1 <_{\mathrm{B}'} b_0$. Suppose $c_1 = 0$. Then $f_{a,b_1} : \mathsf{C} \simeq \mathsf{X} \subseteq \mathsf{C}[c$ for some X giving a contradiction. Hence $0 <_{\mathsf{C}} c$. Suppose $b_1 <_{\mathrm{B}'} b <_{\mathrm{B}'} b_0$. Then there is a $c' <_{\mathsf{C}} c$ such that

$$pj(c', a) = j(0, b).$$

But

$$pj(c, a) = j(0, b_0).$$

Hence by lemma A.5 $b=b_0$ giving a contradiction. Therefore $b_1 \dashv_{B'} b_0$.

The following will be used throughout the rest of the proof.

A.7 DEFINITION. Suppose $b_0 \in B$ such that if $b_0=0$ then $0 \in C_2$. Define recursively a^n, a_n, b^{n+1}, b_{n+1} for all $n \geq 0$ by

$$(*) \qquad \begin{aligned} pj(a^n, a_n) &= j(0, b_n), \\ pj(0, a_n) &= j(b^{n+1}, b_{n+1}). \end{aligned}$$

Also define

$$(**) \qquad \begin{aligned} A_n &= \mathrm{COT}(C[a^n), \\ B_{n+1} &= \mathrm{COT}(C[b^{n+1}), \end{aligned}$$

for each $n \geq 0$.

A.8 LEMMA. 1) If $b_0=0$, $0 \in C_2$, $a_0 \neq 0$, then $a_0 \dashv_{A'} 0$ and $a^0 \neq 0$,

2) if $b_0 \in B'$, ($b_0=0$ implies $0 \in C_2$) then for all $n \geq 0$

$$a^{n+1} \neq 0, \; b^{n+1} \neq 0,$$

$$a_{n+1} \dashv_{A'} a_n, \; b_{n+1} \dashv_{B'} b_n$$

and

$$B_{n+1} + A_n = C, \quad A_{n+1} + B_{n+1} = C.$$

PROOF. 1) Assume the hypothesis. Then $a_0 <_{A'} 0$. Suppose $a_0 <_{A'} a <_{A'} 0$, i.e. $a_0 <_A a$. Then $f_{a,0}^{-1}: C \simeq X \subseteq C_2$ for some X giving a contradiction. So $a_0 \dashv_{A'} 0$. Now suppose $a^0=0$. Then $f_{a_0,0}^{-1}: C \simeq X \subseteq C_2$ for some X, again giving a contradiction. So $a^0 \neq 0$.

2) Assume the hypothesis. By lemma A.6, if $a^n \neq 0$, then $b^{n+1} \neq 0$ and $b_{n+1} \dashv_{B'} b_n$. By the dual of lemma A.6 if $b^{n+1} \neq 0$ then $a^{n+1} \neq 0$ and $a_{n+1} \dashv_{A'} a_n$. But $a^0 \neq 0$. Hence by induction we have

$$a^{n+1} \neq 0, \quad b^{n+1} \neq 0,$$

$$a_{n+1} \dashv_{A'} a_n, \; b_{n+1} \dashv_{B'} b_n,$$

for all $n \geq 0$.

By these results we must have

$$f_{a_{n+1}, b_{n+1}}: C[b^{n+1} \simeq a^{n+1})C$$

and

$$f_{a_n, b_{n+1}}: b^{n+1})C \simeq C[a^n.$$

But $b^{n+1})C$, $C[b^{n+1}$ are separated by the disjoint r.e. sets

$$\delta f_{a_{n+1}, b_{n+1}}, \delta f_{a_n, b_{n+1}}.$$

Hence

$$A_{n+1} + B_{n+1} = C.$$

Also $a^n)\mathsf{C}$, $\mathsf{C}[a^n$ are separated by the disjoint r.e. sets

$$\rho f_{a_n, b_{n+1}}, \rho f_{a_n, b_n} \text{ so that } B_{n+1} + A_n = C.$$

A.9 LEMMA. If $b_0 \in \mathrm{B}$ is not generated then $A_0 + C = C$.

PROOF. Suppose $b_0 \in \mathrm{B}$ is not generated, then $a^0 \neq 0$. From (i), (ii) below it will follow that $f_{a_0, b_0} : \mathsf{C} \simeq a^0)\mathsf{C}$ giving the conclusion of the lemma as we have already observed the r.e. separability of $\mathsf{C}[a^0, a^0)\mathsf{C}$.

(i) If $0 <_{\mathsf{C}} c$ and $f_{a,b_0}(c) = c'$ for $a \in \mathrm{A}'$ then $a = a_0$ and $a^0 <_{\mathsf{C}} c'$.

Clearly $a_0 \leq_{\mathrm{A}'} a$. If $a_0 <_{\mathrm{A}'} a$ then $pj(0, a) = j(c'', b_0)$ for some $c'' \in \mathrm{C}$, i.e. b_0 is generated contradicting the hypothesis of the lemma. So $a = a_0$ and then trivially $a^0{}_{\mathsf{C}} < c'$.

(ii) If $a^0 <_{\mathsf{C}} c'$ and $f_{a_0, b}(c) = c'$ then $b = b_0$ and $0 <_{\mathsf{C}} c$.

This follows by lemma A.5.

A.10 LEMMA. Every $b_0 \in \mathrm{B}$ is generated.

PROOF. Suppose $b_0 \in \mathrm{B}$ is not generated. Then $a^0 \neq 0$ and by lemmata A.8 and A.9

$$(\mathrm{R}_n) \qquad A_n + B_n = C,$$

$$(\mathrm{Q}_n) \qquad B_{n+1} + A_n = C,$$

for all $n \geq 0$, where we have temporarily put $B_0 = C$. But $\mathrm{COT}(\mathsf{C}[b^1) = B_1 < B_0 = C$ as C is a quord. By lemma A.3, (R_n) and (R_{n+1}), $B_{n+1} < B_n$ implies $A_n < A_{n+1}$. By lemma A.3, (Q_n) and (Q_{n+1}) $A_n < A_{n+1}$ implies $B_{n+2} < B_{n+1}$. So by induction $A_n < A_{n+1}$, $B_{n+1} < B_n$ for all $n \geq 0$. But $B_{n+1} < B_n$ implies $b^{n+1} <_{\mathsf{C}} b^n$. Also $\{b^n\}_{n>0}$ is a recursive sequence. This contradicts the fact that C is quasi-well-ordered and we conclude that b_0 is generated.

A.11 LEMMA. If $b_0 \in \mathrm{B}$ and $a^0 \neq 0$ then there is an $a_{-1} \in \mathrm{A}'$ such that $a_0 \dashv_{\mathrm{A}'} a_{-1}$ and a b^0 such that

$$(***) \qquad pj(0, a_{-1}) = j(b^0, b_0).$$

PROOF. By lemma A.10, b_0 is generated, i.e. there is an $a_{-1} \in \mathrm{A}'$, $b^0 \in \mathrm{C}$ such that (***) holds. Then by lemma A.6, $a_0 \dashv_{\mathrm{A}'} a_{-1}$ and $b^0 \neq 0$.

Call $a \in \mathrm{A}$ a *tail element* of A if there is a finite sequence

$$a = a_0 \dashv_{\mathsf{A}} a_1 \dashv_{\mathsf{A}} \cdots \dashv_{\mathsf{A}} a_n \dashv_{\mathsf{A}'} 0 .$$

Dually we define the tail elements of B.

A.12 LEMMA. Let $b_0 \in \mathrm{B}$ and $a^0 \neq 0$ be such that if b_0 is a tail element of B then $p : \mathsf{CA} \simeq \mathsf{CB}$. Then there are a_n, b_n for $n<0$ such that

$$pj(0, a_n) = j(b^1, b_{n+1}),$$
$$pj(a^0, a_n) = j(0, b_n)$$

and $a_{n+1} \dashv_{\mathsf{A}} a_n$, $b_{n+1} \dashv_{\mathsf{B}} b_n$ for all $n (n \geq 0$ or $n<0)$.

PROOF. Assume the hypothesis. We have already defined a^n, a_n, b^{n+1}, b_{n+1} for $n \geq 0$ satisfying (*). By lemma A.10 there are a_{-1} and $b^0 \neq 0$ such that

$$pj(0, a_{-1}) = j(b^0, b_0) \quad \text{and} \quad a_0 \dashv_{\mathsf{A}'} a_{-1} .$$

If $a_{-1} = 0$ then we must have $b_0 \dashv_{\mathsf{B}'} 0$ so that b_0 is a tail element of B and by the hypothesis of the lemma

$$p : \mathsf{CA} \simeq \mathsf{CB} .$$

But $pj(0, 0) = j(b^0, b_0) \in \mathrm{C}^{\mathrm{c}}\mathsf{CB}$ and $j(0, 0) \notin \mathrm{C}^{\mathrm{c}}\mathsf{CA}$.
Thus $a_{-1} \neq 0$, i.e. $a_{-1} \in \mathrm{A}$.

By the dual of lemma A.11, there are $a^{-1} \neq 0$ and b_{-1} such that

$$pj(a^{-1}, a_{-1}) = j(0, b_{-1}) \quad \text{and} \quad b_0 \dashv_{\mathsf{B}'} b_{-1}$$

and as above we may show that $b_{-1} \neq 0$ so that $b_{-1} \in \mathrm{B}$.

Repeating the above procedure we construct a^n, a_n, b^{n+1}, b_n for $n<0$ satisfying (*) such that

$$a_{n+1} \dashv_{\mathsf{A}} a_n \quad \text{and} \quad b_{n+1} \dashv_{\mathsf{B}} b_n .$$

Define A_n, B_{n+1} for $n<0$ as in (**); then as in lemma A.8 we have that

$$(\mathrm{R}_n) \qquad A_{n+1} + B_{n+1} = C ,$$
$$(\mathrm{Q}_n) \qquad B_{n+1} + A_n = C ,$$

for $n<0$ as well as for $n \geq 0$.
To conclude the proof of the lemma we show that $a^{n+1} = a^n$ and $b^{n+1} = b^n$ for all n, so that $a^n = a^0$, $b^n = b^1$ for all n.

Suppose $a^{n+1} \neq a^n$. Then $a^n <_{\mathsf{C}} a^{n+1}$ or $a^{n+1} <_{\mathsf{C}} a^n$. If $a^n <_{\mathsf{C}} a^{n+1}$ then $A_n < A_{n+1}$ and as in lemma A.10 $B_{m+1} < B_m$ for all $m>n$ giving a contradiction. If $a^{n+1} <_{\mathrm{C}} a^n$ then $A_{n+1} < A_n$ and as in lemma A.10 $A_{m+1} < A_m$ for all $m \geq n$ again giving a contradiction. Hence $a^{n+1} = a^n$. So $A_n = A_{n+1}$ which by R_n, R_{n+1} and corollary 3.2.9 gives $B_n = B_{n+1}$. Hence $b^n = b^{n+1}$ as B_n is a quord, concluding the proof of the lemma.

From lemma A.12 it is immediately clear that if b_0 is a tail element of B and $p : \mathsf{CA} \simeq \mathsf{CB}$ then $a^0 = 0$.

In order to complete the proof of the theorem we consider the following exclusive and exhaustive cases:

1 $p : \mathsf{CA} \simeq \mathsf{CB}$,

2 $pj(c, a) = j(0, 0)$ for some $c \in \mathrm{C}$, $a \in \mathrm{A}$,

3 $pj(0, 0) = j(c, b)$ for some $c \in \mathrm{C}$, $b \in \mathrm{B}$.

Case 1. $p : \mathsf{CA} \simeq \mathsf{CB}$.

Let $q(a) = b$ if, and only if

$$pj(0, a) = j(b^1, b),$$
$$pj(a^0, a_0) = j(0, b),$$
$$pj(0, a_0) = j(b^1, b_1),$$

for some a^0, a_0, b^1, b_1.

Clearly q is a one-one partial recursive sub-function of q_1. By lemma A.10 and the assumption that $p : \mathsf{CA} \simeq \mathsf{CB}$ it follows that q_1 maps A onto B. Hence to show that $q : \mathsf{A} \simeq \mathsf{B}$ we need only prove that q is defined on A.

Let $a \in \mathrm{A}$, $pj(0, a) = j(b^1, b)$. If $b^1 = 0$, then clearly $q(a) = b$ with $a^0 = b^1 = 0$ and $a_0 = a$, $b_1 = b$. If $b^1 \neq 0$, let $pj(a^0, a_0) = j(0, b)$, then $a^0 \neq 0$ and the hypothesis of lemma A.12 holds with $b_0 = b$. Hence by lemma A.12, $pj(0, a_0) = j(b^1, b_1)$ so that $q(a) = b$.

Case 2. $pj(c, a) = j(0, 0)$ for $c \in \mathrm{C}$, $a \in \mathrm{A}$.

Clearly we must have $0 \in \mathrm{C}_2$. Hence the hypotheses of lemma A.8 1) and 2) hold with $b_0 = 0$, $a_0 = a$, $a^0 = c$. Define the recursive functions r_1, r_2 by $r_1(n) = a_n$, $r_2(n) = b_{n+1}$ for all $n \geq 0$. Then by lemma A.8

$$r_1(n+1) \dashv_{\mathsf{A}} r_1(n), \quad r_1(0) \dashv_{\mathsf{A}'} 0,$$
$$r_2(n+1) \dashv_{\mathsf{B}} r_2(n), \quad r_2(0) \dashv_{\mathsf{B}'} 0$$
$$\text{and} \quad q_1(r_1(n)) = r_2(n).$$

Let $q(a) = b$ if, and only if,

$$\text{either} \begin{cases} pj(0, a) = j(b^1, b), \\ pj(a^0, a_0) = j(0, b), \\ pj(0, a_0) = j(b^1, b_1) \quad \text{for some } a^0, a_0, b^1, b_1, \end{cases}$$

or there is an n such that $a=r_1(n)$ and $b=r_2(n)$.

Clearly q is a one-one partial recursive sub-function of q_1 and q_1 maps A onto B. It remains to show that q maps A onto B.

Let $a \in$A. If a is a tail element of A then $a=r_1(n)$ for some n and $q(a)=r_2(n)$.

If a is not a tail element of A and $pj(0, a)=j(b^1, b)$ then $b^1=0$ implies $q(a)=b$. If $b^1 \neq 0$ then if $pj(a^0, a_0)=j(0, b)$ holds so does the hypothesis of lemma A.12 with $b_0=b$ since b cannot be a tail element of B. Hence by lemma A.12 $pj(0, a_0)=j(b^1, b_1)$ so that $q(a)=b$.

Case 3. $pj(0, 0)=j(c, b)$ for $c \in$C, $b \in$B. Using the dual argument to case 2 we have a sub-function q of q_2 such that q: B ~ A.

A.13 THEOREM. If $C \geq 1$ is a quord, then P_C is an order embedding, i.e.

1) $CA=CB$ if, and only if $A=B$,

2) $(\exists D)(CA+D=CB)$ if, and only if $(\exists D)(A+D=B)$.

PROOF. 1) follows from theorem A.4 with $C_1=C_2=0$.

2) The implication from right to left follows from the proof of theorem 6.2.1.

Now suppose $CA+D=CB$ for some D, then by lemma A.14 below, $CA=CB'+C_2$ for some B', C, where $(\exists E)(B'+E=B)$ and $(\exists F)(C_2+F=C \,\&\, F \neq 0)$. Hence by theorem A.4, $A=B'$ and the required result follows.

A.14 LEMMA. If $C \geq 1$ is a quord, then $(\exists E)(D+E=CB)$ implies $D=CB'+C_2$ for some B', C_2 such that $(\exists F)(B'+F=B)$ and $(\exists G)(C_2+G=C \,\&\, G \neq 0)$.

PROOF (generalizing lemma 9.2.2). Let C $\in C$, B $\in B$, and $D+E=CB$. Choose D $\in D$ to be an initial segment of CB. Let C=C'C, B=C'B, D=C'D, and let

$$\mathsf{B}' = \mathsf{B}\left[\{b:(\forall c)\left(c \in \mathsf{C} \Rightarrow j(c, b) \in \mathsf{D}\right)\}\right].$$

Let B′=C‘B′.

As D is recursively separable from C‘CB − D we must have that B′ is recursively separable from B − B′. Hence, as B′ is an initial segment of B, B'=COT(B′) is such that $B'+F=B$ for some C.O.T. F.

If D ≠ CB′ then there is a (unique) $b \in$B − B′ such that $j(c, b) \in$D for some $c \in$C. Let $\mathsf{C}_2=\mathsf{C}[\{c:j(c, b) \in \mathsf{D}\}$. Then C_2 is a proper initial segment of C and $\mathsf{C}`\mathsf{C}_2$ is recursively separable from $\mathsf{C}-\mathsf{C}`\mathsf{C}_2$ so that $C_2=\mathrm{COT}(\mathsf{C}_2)$ is such that $C_2+G=C$ for some non-zero G, since C is a quord. Also

$$\mathsf{D}=\mathsf{C}\cdot\mathsf{B}' \mathbin{\hat{+}} \mathsf{C}_2\cdot\{\langle b, b\rangle\},$$

hence $D=CB'+C_2$.

APPENDIX B

INFINITE PRODUCTS AND PRINCIPAL NUMBERS FOR MULTIPLICATION

by A. G. HAMILTON

We first of all define infinite products of sequence orderings in a manner closely related to the definition of exponentiation and then proceed to show that there exist $2^{\aleph_0}$ principal numbers for multiplication of classical ordinal ω^{ω^m} which are not of the form W^A.

B.1 DEFINITION. Let $A = \langle \mathsf{A}_i : \mathsf{C} \rangle$ be a standard sequence ordering (see following definition 12.2.3) such that every A_i has a minimum element. Let

$$E(A) = \{K : K = \begin{pmatrix} c_0 \dots c_n \\ a_0 \dots a_n \end{pmatrix} \;\&\; c_0 >_{\mathsf{C}} c_1 \dots >_{\mathsf{C}} c_n$$
$$\& \; (\forall m)\,(0 \le m \le n \Rightarrow a_m \in \mathsf{C}`\mathsf{A}_{c_m} \;\&\; a_m \neq \min \mathsf{A}_{c_m})\}\,.$$

(K may be the empty bracket symbol.) Now define

$$\Pi(A) = \{\langle e(K), e(K')\rangle : K, K' \in E(A)$$
$$\& \; K = \begin{pmatrix} c_0 \dots c_m \\ a_0 \dots a_m \end{pmatrix} \;\&\; K' = \begin{pmatrix} c'_0 \dots c'_n \\ a'_0 \dots a'_n \end{pmatrix}$$
$$\& \; [K = \mathbf{0} \vee (K \neq 0 \;\&\; [(m \le n \;\&\; (\forall r)\,(r \le m \Rightarrow c_r = c'_r \;\&\; a_r = a'_r))$$
$$\vee\, (\exists r)\,(\forall s)\,\{(s < r \Rightarrow a_s = a'_s \;\&\; c_s = c'_s) \;\&\; (\langle c_r, c'_r\rangle \in \mathsf{C} \;\&\; c_r \neq c'_r$$
$$.\vee.\; c_r = c'_r \;\&\; \langle a_r, a'_r\rangle \in \mathsf{A}_{c_r} \;\&\; a_r \neq a'_r)\}])]\}\,.$$

B.2 LEMMA. If $\langle \mathsf{A}_i : \mathsf{C} \rangle$ is a standard sequence ordering with $\mathsf{C}`\mathsf{C}$ r.e. and C has a minimum element, c_0, say, then

$$\Pi\langle \mathsf{A}_i : \mathsf{C}\rangle \simeq \mathsf{A}_{c_0} \cdot \Pi\langle \mathsf{A}_i : \mathsf{C}[(\mathsf{C}`\mathsf{C} - \{c_0\})\rangle\,.$$

PROOF. By our definition of standard sequence ordering every $\mathsf{A}_i \subseteq \mathsf{R}$ (and $\mathsf{C} \subseteq \mathsf{R}$). Let f be the function defined by

$$fe\begin{pmatrix} c_1 \dots c_n \\ a_1 \dots a_n \end{pmatrix} = \begin{cases} j\left(a_n, e\begin{pmatrix} c_1 \dots c_{n-1} \\ a_1 \dots a_{n-1} \end{pmatrix}\right) & \text{if } c_n = c_0 \text{ \& each } c_i \in \mathsf{C}^{\text{‘}}\mathsf{C}, \\ j\left(\min(\mathsf{A}_{c_0}), e\begin{pmatrix} c_1 \dots c_n \\ a_1 \dots a_n \end{pmatrix}\right) & \text{if } c_n \neq c_0 \\ & \text{\& each } c_i \in \mathsf{C}^{\text{‘}}\mathsf{C}, \\ \text{undefined if any } c_i \notin \mathsf{C}^{\text{‘}}\mathsf{C} \text{ or if the argument} \\ \text{of } f \text{ is not of the above form.} \end{cases}$$

It is easily verified that f is one-one and maps $\mathsf{C}^{\text{‘}}\Pi\langle \mathsf{A}_i : \mathsf{C}\rangle$ onto $\mathsf{C}^{\text{‘}}\mathsf{A}_{c_0} \cdot \Pi\langle \mathsf{A}_i : \mathsf{C}[(\mathsf{C}^{\text{‘}}\mathsf{C} - \{c_0\})\rangle$. Since, by hypothesis $\mathsf{C}^{\text{‘}}\mathsf{C}$ is r.e. and $\mathrm{Seq} = \mathsf{C}^{\text{‘}}\mathsf{R}$ is r.e. it follows that δf is r.e. Finally we leave the reader to check that f is order preserving on the whole of its domain.

B.3 COROLLARY. With the hypotheses of lemma B.2, if $\mathsf{D} \subseteq \mathsf{C}$ and $\min(\mathsf{D}) = c_0$ then

$$\Pi\langle \mathsf{A}_i : \mathsf{D}\rangle \simeq \mathsf{A}_{c_0} \cdot \Pi\langle \mathsf{A}_i : \mathsf{D}[(\mathsf{C}^{\text{‘}}\mathsf{D} - \{c_0\})\rangle.$$

PROOF. The same function f as in the proof of the lemma is the required recursive isomorphism.

Notation: If C has a first element, we shall denote $\mathsf{C}[(\mathsf{C}^{\text{‘}}\mathsf{C} - \{\min(\mathsf{C})\})$ by C'. If C has an initial segment of type n (finite) and $n > 1$, then $\mathsf{C}^{(n)}$ denotes $(\mathsf{C}^{(n-1)})'$.

B.4 COROLLARY. Let C be a well-ordering of type ω ($\subseteq \mathsf{R}$) with r.e. field, and let D be a sub-ordering of C with an initial segment $d_0, \dots, d_{n-1}$ (i.e. the first n elements are $d_0 \dots$ in that order). Then $\Pi\langle \mathsf{A}_i : \mathsf{D}\rangle \simeq \mathsf{A}_{d_0} \cdot \dots \cdot \mathsf{A}_{d_{n-1}} \cdot \Pi\langle \mathsf{A}_i : \mathsf{D}^{(n)}\rangle$.

PROOF. If $n = 1$, let $\mathsf{C}_1 = \mathsf{C}[\{x : x \in \mathsf{C}^{\text{‘}}\mathsf{C} \,\&\, \langle d_0, x\rangle \in \mathsf{C}\}$. Since d_0 can have at most a finite number of predecessors in C, C_1 has an r.e. field and we can apply corollary B.3 to get

$$\Pi\langle \mathsf{A}_i : \mathsf{D}\rangle \simeq \mathsf{A}_{d_0} \cdot \Pi\langle \mathsf{A}_i : \mathsf{D}'\rangle.$$

Now assume $n > 1$. Suppose, as induction hypothesis, that

$$\Pi\langle \mathsf{A}_i : \mathsf{D}\rangle \simeq \mathsf{A}_{d_0} \cdot \dots \cdot \mathsf{A}_{d_{n-2}} \cdot \Pi\langle \mathsf{A}_i : \mathsf{D}^{(n-1)}\rangle.$$

We have to show that

$$\Pi\langle \mathsf{A}_i : \mathsf{D}^{(n-1)}\rangle \simeq \mathsf{A}_{d_{n-1}} \cdot \Pi\langle \mathsf{A}_i : \mathsf{D}^{(n)}\rangle .$$

But this follows immediately from the first part of the proof. So we conclude

$$\Pi\langle \mathsf{A}_i : \mathsf{D}\rangle \simeq \mathsf{A}_{d_0} \cdot \dots \cdot \mathsf{A}_{d_{n-1}} \cdot \Pi\langle \mathsf{A}_i : \mathsf{D}^{(n)}\rangle .$$

We now define a collection $\mathscr{U}$ of sub-orderings U of W' as follows:

By theorem B.5 below, there exists a collection of $2^{\aleph_0}$ isolated sets none of which is mapped into any other by a finite-to-one partial recursive function. We can suppose that none of these sets contains 0, 1 or 2. Now order each set by magnitude, and we get $2^{\aleph_0}$ well-orderings of type ω. For each such well-ordering U' let $\mathsf{U} = \{\langle\langle m\rangle, \langle n\rangle\rangle : \langle m, n\rangle \in \mathsf{U}'\}$. Then the collection $\mathscr{U}$ of all such U has $2^{\aleph_0}$ members, each of which is of type ω, $\subseteq \mathsf{W}'$ but not recursively isomorphic to W' and moreover no two members of $\mathscr{U}$ are recursively isomorphic and no finite-to-one partial recursive function maps the field of one into the field of any other.

Every $\mathsf{U} \in \mathscr{U}$ satisfies the hypotheses of corollary B.4 which apply to D (with W' as the well-ordering C).

B.5 THEOREM. There exists a collection $\mathscr{A} = \{\mathrm{A}_i : i \in \mathrm{I}\}$ of sets of (non-negative) integers with the following properties:

(i) the cardinal of I is $2^{\aleph_0}$,

(ii) for all $i, j \in \mathrm{I}$ with $i \neq j$ there is no finite-to-one partial recursive function mapping A_i into A_j,

(iii) each A_i $(i \in \mathrm{I})$ is isolated.

PROOF. Let $\varphi_0, \varphi_1, \dots$ be an enumeration (non-effective) of all finite-to-one partial recursive functions with infinite range.

We construct a tree

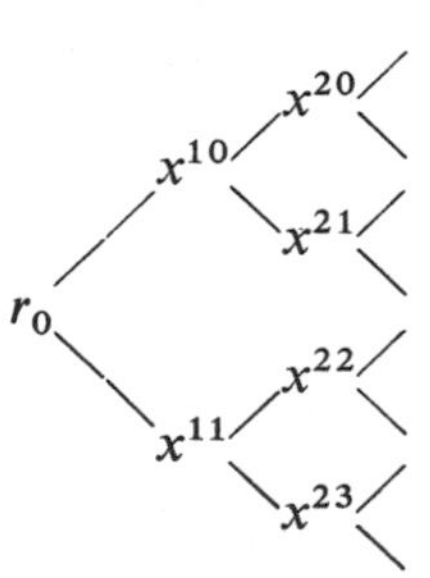

which branches into two at each node, each node being an integer. The sets A_i will be the branches of the completed tree.

Stage 0. Let r_0 be any integer not equal to $\varphi_0(z_0)$ where

$$z_0 = \nu_w\{\varphi_0(w) \text{ is defined}\}.$$

We describe stage s for $s \geq 1$.
Stage s. We define successively $x^{s0}, x^{s1}, \ldots, x^{s\,2^s-1}$.
Define x^{si} for $0 \leq i \leq 2^s - 1$ as follows:

Let T_{si} = the set of all elements of the tree which have been defined before x^{si},

B_{si} = the set of all elements defined before x^{si} which lie on the same branch as x^{si},

T'_s = the set of all elements of the tree defined when stage $s-1$ has been completed.

Choose x^{si} to be some (say the least) integer t satisfying

(a) $t \notin T_{si}$,

(b) either $t \notin \bigcup_{j \leq s} \delta\varphi_j$ or $t \in \bigcup_{j \leq s} \delta\varphi_j$ and $(\forall j \leq s)\,(t \in \delta\varphi_j \Rightarrow \varphi_j(t) \notin T_{si})$,

(c) $(\forall j \leq s)\,(\forall z)\,(z \in T_{si} \cap \delta\varphi_j \Rightarrow \varphi_j(z) \neq t)$,

(d) $t \notin \{\varphi_j(z_j) : 0 \leq j \leq s+1\}$, where $z_j = \nu_w\{\varphi_j(w)$ is defined and $\varphi_j(w) \notin \{\varphi_k(z_k) \mid k < j\} \cup T'_j\}$.

At every stage we have infinitely many numbers to choose from; each of conditions (a)–(d) disqualifies only a finite number of integers since each j is finite-to-one with infinite range. Thus such a t always exists.

This describes the construction of the tree. The branches of this tree are the sets A_i.

(i) The cardinal of I is $2^{\aleph_0}$ since all the branches are distinct,

(ii) Suppose $\varphi_k : A_i \xrightarrow{\text{into}} A_j$ where $i \neq j$. Then $A_i \neq A_j$ by the construction, and we may find $x \in A_i - A_j$ such that x appears in the construction of the tree after the kth stage. Then $y = \varphi_k(x) \in A_j$ and $y \neq x$. If y appears in the tree before x, then (b) is violated when x is put in the tree. If y appears after x then (c) is violated when y is put in. Thus we have a contradiction so condition (ii) of the theorem is satisfied.

(iii) Suppose A_i is not isolated, then A_i contains an infinite r.e. set, say $\rho\varphi_k$ (it must be the range of one of the φ_j's). At stage k we kept $\varphi_k(z_k)$ out of every branch of the tree, and at all later stages. But z_k was chosen so that $\varphi_k(z_k)$ did not occur in the tree before stage k. So $\varphi_k(z_k) \notin A_i$ which is a contradiction. Thus condition (iii) of the theorem is met and the theorem is proved.

We now write u_i for the ith element of U in the obvious way.

B.6 LEMMA. If $\mathsf{U} \in \mathscr{U}$, $\{P_i\}_{i<\omega}$ is a strictly increasing sequence of principal numbers for multiplication and $\mathsf{P}_i \in P_i$ for all i, then $\mathrm{COT}\,\Pi\langle \mathsf{P}_i\colon \mathsf{U}\rangle$ is a principal number for multiplication.

PROOF. We have to show that if B is recursively isomorphic to an initial segment of $\Pi\langle \mathsf{P}_i\colon \mathsf{U}\rangle$ then

$$\mathsf{B}\cdot\Pi\langle \mathsf{P}_i\colon \mathsf{U}\rangle \simeq \Pi\langle \mathsf{P}_i\colon \mathsf{U}\rangle .$$

Suppose

$$\mathsf{B} \simeq \mathsf{B}^1 < \Pi\langle \mathsf{P}_i\colon \mathsf{U}\rangle$$

and suppose

$$\mathsf{U} = \{\langle u_i, u_j\rangle : i \leq j\} .$$

We must have

$$\mathsf{B}^1 < \Pi\left\langle \mathsf{P}_i\colon \mathsf{U}\left[e\begin{pmatrix} u_j \\ \min \mathsf{P}_{u_j}\end{pmatrix}\right)\right\rangle \quad \text{for some } j ,$$

i.e.

$$\begin{aligned} \mathsf{B}^1 &< \Pi\langle \mathsf{P}_i\colon \mathsf{U}[u_j\rangle \\ &= \mathsf{P}_{u_0}\cdot\ldots\cdot\mathsf{P}_{u_{j-1}} \\ &\sim \mathsf{P}_{u_{j-1}}, \end{aligned}$$

since $P_{u_{j-1}}$ is a principal number for multiplication $> P_{u_i}$ for $i<j-1$. Now, using corollary B.4 and properties of principal numbers we have:

$$\begin{aligned} \mathsf{B}\cdot\Pi\langle \mathsf{P}_i\colon \mathsf{U}\rangle &\simeq \mathsf{B}^1\cdot\Pi\langle \mathsf{P}_i\colon \mathsf{U}\rangle \\ &\simeq \mathsf{B}^1\cdot\mathsf{P}_{u_0}\cdot\mathsf{P}_{u_1}\cdot\ldots\cdot\mathsf{P}_{u_{j-1}}\cdot\Pi\langle \mathsf{P}_i\colon \mathsf{U}^{(j)}\rangle \\ &\simeq \mathsf{B}^1\cdot\mathsf{P}_{u_{j-1}}\cdot\Pi\langle \mathsf{P}_i\colon \mathsf{U}^{(j)}\rangle \\ &\simeq \mathsf{P}_{u_{j-1}}\cdot\Pi\langle \mathsf{P}_i\colon \mathsf{U}^{(j)}\rangle \\ &\simeq \mathsf{P}_{u_0}\cdot\ldots\cdot\mathsf{P}_{u_{j-1}}\cdot\Pi\langle \mathsf{P}_i\colon \mathsf{U}^{(j)}\rangle \\ &\simeq \Pi\langle \mathsf{P}_i\colon \mathsf{U}\rangle . \end{aligned}$$

Hence $\mathrm{COT}\,\Pi\langle \mathsf{P}_i\colon \mathsf{U}\rangle$ is a principal number for multiplication.

B.7 LEMMA. If $\mathsf{B} \simeq \mathsf{W}^{\mathsf{A}}$ for some A, then the function $\lambda\alpha\alpha 2$ is recursively representable on B.

PROOF. If $\mathsf{B} = \mathsf{W}^{\mathsf{A}}$ then $\mathrm{C}`\mathsf{B}$ consists of elements of the form

$$e\begin{pmatrix} a_0 \ldots a_m \\ n_0 \ldots n_m \end{pmatrix} \quad \text{where} \quad a_0 >_{\mathsf{A}} \cdots >_{\mathsf{A}} a_m$$

and the n_i $(0 \leq i \leq m)$ are non-zero integers.

The function f defined on ρe as follows recursively represents the function $\lambda\alpha\alpha 2$ on B:

$$fe\begin{pmatrix} a_0 \dots a_m \\ n_0 \dots n_m \end{pmatrix} = e\begin{pmatrix} a_0\, a_1 \dots a_m \\ 2n_0\, n_1 \dots n_m \end{pmatrix},$$

i.e.

$$|f(x)|_{\mathsf{B}} = |x|_{\mathsf{B}} \cdot 2 \quad \text{whenever} \quad x \in \mathrm{C}`\mathsf{B}$$

and

$$f(x) \notin \mathrm{C}`\mathsf{B} \quad \text{if} \quad x \notin \mathrm{C}`\mathsf{B}.$$

This function f is obviously one-one and partial recursive mapping $\mathrm{C}`\mathsf{B}$ into $\mathrm{C}`\mathsf{B}$. That it represents $\lambda\alpha\alpha 2$ on B can be seen by mapping the set

$$\left\{ x : \left\langle e\begin{pmatrix} a_0 \dots a_m \\ n_0 \dots n_m \end{pmatrix}, x \right\rangle \in \mathsf{B} \;\&\; \left\langle x, e\begin{pmatrix} a_0 a_1 \dots a_m \\ 2n_0 n_1 \dots n_m \end{pmatrix} \right\rangle \in \mathsf{B} \right\},$$

in a one-one order preserving way onto the set

$$\left\{ y : \left\langle y, e\begin{pmatrix} a_0 \dots a_m \\ n_0 \dots n_m \end{pmatrix} \right\rangle \in \mathsf{B} \right\},$$

as follows

$$e\begin{pmatrix} a_0\, a_1' \dots a_1' \\ n\, n_1' \dots n_1' \end{pmatrix} \to e\begin{pmatrix} a_0\, a_1' \dots a_1' \\ n - n_0\, n_1' \dots n_1' \end{pmatrix} \quad \text{if } n > n_0,$$

$$e\begin{pmatrix} a_1' \dots a_1' \\ n_1' \dots n_1' \end{pmatrix} \quad \text{if } n = n_0.$$

Now if $g: \mathsf{B}^1 \simeq \mathsf{B} = \mathsf{W}^{\mathsf{A}}$ then $g^{-1}fg$ recursively represents $\lambda\alpha\alpha 2$ on B^1.

B.8 Lemma. If $\mathsf{U} \in \mathscr{U}$ and $\{P_i\}_{i<\omega}$ is a strictly increasing sequence of principal numbers for multiplication, then $\mathsf{P}_i \in P_i$ can be chosen so that the function $\lambda\alpha\alpha 2$ is not recursively representable on $\mathbf{\Pi}\langle \mathsf{P}_i : \mathsf{U} \rangle$.

Proof. Let $\mathsf{U} = \{\langle u_i, u_j \rangle : i \leq j\}$. Since all the P_i are principal numbers, we must have $P_i > 3$ for all i, so we can choose the $\mathsf{P}_i \in P_i$ with the following properties:

$$\min(\mathsf{P}_i) = 0, \min(\mathsf{P}_i') = \langle 1 \rangle = 2, \min(\mathsf{P}_i^{(2)}) = u_i.$$

That we can choose orderings thus and still have them included in R is easily seen (e.g. if $\mathsf{Q}_i \in P_i$, take $\widehat{u_i + 2\mathsf{Q}_i}$ and replace the first three elements by 0, 2, u_i. The order must remain correct since $u_i > 2$ for all i.)

With this choice of $\{\mathsf{P}_i\}_{i<\omega}$, consider

$$e\begin{pmatrix} j \\ 2 \end{pmatrix} \in \mathrm{C}`\mathbf{\Pi}\langle \mathsf{P}_i : \mathsf{U} \rangle.$$

Suppose α is the ordinal represented by $e\binom{j}{2}$ in $\Pi\langle \mathsf{P}_i : \mathsf{U}\rangle$.

Then the element representing $\alpha 2$ is $e\binom{j}{u_j}$. The set of predecessors of $e\binom{j}{2}$ can be mapped exactly twice in a one-one order preserving way into the set of predecessors of $e\binom{j}{u_j}$ by the identity function and the function

$$e\begin{pmatrix} c_0 \dots c_n \\ a_0 \dots a_n \end{pmatrix} \to e\begin{pmatrix} j\, c_0 \dots c_n \\ 2\, a_0 \dots a_n \end{pmatrix}.$$

Now if $\lambda\alpha\alpha 2$ were recursively representable on $\Pi\langle \mathsf{P}_i : \mathsf{U}\rangle$, then the function defined on numbers of the form $e\binom{j}{2}$, taking $e\binom{j}{2}$ to $e\binom{j}{u_j}$ would be partial recursive. However, this would imply that $\lambda i u_i$ was a (partial) recursive one-one function. But U is not recursively isomorphic to W, so $\lambda i u_i$ is not a recursive function. Therefore $\lambda\alpha\alpha 2$ is not recursively representable on $\Pi\langle \mathsf{P}_i : \mathsf{U}\rangle$.

B.9 COROLLARY. $\mathrm{COT}\,\Pi\langle \mathsf{P}_i : \mathsf{U}\rangle$ is a principal number for multiplication not of the form W^A.

PROOF. By lemma B.6, $\mathrm{COT}\,\Pi\langle \mathsf{P}_i : \mathsf{U}\rangle$ is a principal number for multiplication. Suppose it were of the form W^A, then

$$\Pi\langle \mathsf{P}_i : \mathsf{U}\rangle \simeq \mathsf{W}^{\mathsf{A}} \quad \text{for some } \mathsf{A}. \tag{*}$$

By lemma B.8 $\lambda\alpha\alpha 2$ is not recursively representable on $\Pi\langle \mathsf{P}_i : \mathsf{U}\rangle$. By lemma B.7, $\lambda\alpha\alpha 2$ is recursively representable on W^{A}. This contradicts (*) and the corollary is proved.

B.10 LEMMA. If $\mathsf{U}, \mathsf{V} \in \mathscr{U}$ and $\mathsf{U} \neq \mathsf{V}$ then

$$\mathrm{COT}\,\Pi\langle \mathsf{P}_i : \mathsf{U}\rangle \neq \mathrm{COT}\,\Pi\langle \mathsf{P}_i : \mathsf{V}\rangle.$$

PROOF. Suppose the contrary, then there is a partial recursive one-one function f such that

$$f : \Pi\langle \mathsf{P}_i : \mathsf{U}\rangle \simeq \Pi\langle \mathsf{P}_i : \mathsf{V}\rangle.$$

Now

$$fe\binom{u}{2} = e\begin{pmatrix} c_0 \dots c_n \\ a_0 \dots a_n \end{pmatrix} \quad \text{for some} \quad c_0, \dots, c_n \in C\text{`}\mathsf{V} \quad \text{and} \quad a_i \in C\text{`}\mathsf{P}_{c_i}.$$

So

$$l\left(\left(fe\binom{u}{2}\right)_0\right)\in \text{C‘V}$$

and therefore

$$\lambda x l\left(\left(fe\binom{x}{2}\right)_0\right)$$

is a finite-to-one partial recursive function mapping C‘U into C‘V. For suppose there were a $v_r \in$ C‘V which was the image of infinitely many $u_i \in$ C‘U. It would then follow that all of $\Pi\langle \mathsf{P}_i : \mathsf{U}\rangle$ is mapped onto a proper initial segment of $\Pi\langle \mathsf{P}_i : \mathsf{V}\rangle$ by f, namely that initial segment determined by $e\binom{v_{r+1}}{2}$ (since U has type ω and the set of u_i mapped to v_r is cofinal with U). This is a contradiction, so $\lambda x l((fe\binom{x}{2})_0)$ is finite-to-one. That it is partial recursive is obvious.

But the existence of such a function contradicts the definition of $\mathscr{U}$. Hence there is no such f and therefore

$$\text{COT}\,\Pi\langle \mathsf{P}_i : \mathsf{U}\rangle \neq \text{COT}\,\Pi\langle \mathsf{P}_i : \mathsf{V}\rangle .$$

B.11 THEOREM. There exist $2^{\aleph_0}$ principal numbers for multiplication of classical ordinal ω^{ω^ω} which are not of the form W^A.

PROOF. By corollary B.9 $\text{COT}\,\Pi\langle \mathsf{P}_i : \mathsf{U}\rangle$ is a principal number for multiplication not of the form W^A, for each $\mathsf{U}\in\mathscr{U}$. By lemma B.10 all the co-ordinals $\text{COT}\,\Pi\langle \mathsf{P}_i : \mathsf{U}\rangle$ for $\mathsf{U}\in\mathscr{U}$ are distinct. The cardinality of $\mathscr{U}$ is $2^{\aleph_0}$ so there are $2^{\aleph_0}$ principal numbers for multiplication not of the form W^A. But $|\mathsf{U}|=\omega$ for all $\mathsf{U}\in\mathscr{U}$ and the $|\mathsf{P}_i|$ form a strictly increasing ω-sequence of (classical) principal numbers for multiplication. Taking the $|\mathsf{P}_i|$ to be as small as possible implies that the ordinals $|\mathsf{P}_i|$ must form an increasing sequence $\{\omega^{\omega^n}\}_{n<\omega}$. Hence $\text{COT}\,\Pi\langle \mathsf{P}_i : \mathsf{U}\rangle = \prod_{n<\omega} \omega^{\omega^n}$ (classical product) $= \omega^{\omega^\omega}$. This completes the proof.

Theorem 10.2.10 shows that the result in theorem B.11 is best possible.

NOTATION AND TERMINOLOGY

1 Logical notation.

We use logical symbolism freely for convenience in our informal exposition. We write &, $\vee$, $\neg$, $\Rightarrow$, $\Leftrightarrow$, $\exists$, $\forall$, E!, μ_x for and, or, not, implies, if and only if, there exists, for all, there is a unique, the least x such that, respectively. In this last we always mean the least natural number. We occasionally write

$$(\forall x < y) \quad \text{or} \quad (\forall x)_{<y} \quad \text{for} \quad (\forall x)\,(x < y \Rightarrow)$$

and analogously for $\exists$. We also use the λ-notation for functions (see e.g. KLEENE, 1952, p. 34) and we sometimes use dots for bracketing purposes in the usual way but we do not need rules for association since we do not use dots in such complicated ways.

2 Set-theoretic notions and notations.

$\mathcal{N}$ denotes the set of natural numbers otherwise called non-negative integers. We use $\leq$ for the usual ordering of natural numbers by magnitude. p_i denotes the ith prime ($p_0 = 2$) and $(x)_i$ denotes the exponent of the highest power of p_i which divides x. If p_i is the largest prime which divides x non-trivially then we set

$lh(x) = i+1$. N.B. This is a deviation from KLEENE (1952).

As usual the membership relation is denoted by $\in$ and ε is used to denote an ordinal. We write $x \notin A$ for $\neg x \in A$ as usual and similarly for $\neq$. A *set* generally means a set of natural numbers unless we specifically say e.g. "a set of ordinals". We use *class* and *collection* for talking about aggregates of sets or classes. The empty set (of anything) is denoted by $\emptyset$. We use Roman letters A, B, C, ... for sets. $\{x: \mathfrak{P}(x)\}$ denotes the set of all elements satisfying $\mathfrak{P}$. (The context will make clear what kind of elements is intended.) $\{x\}$ denotes the set whose only element is x but $\{a(n)\}$, $\{a_n\}$, $\{a(n)\}_{n=0}^{\infty}$, $\{a_n\}_{n=0}^{\infty}$ all denote sequences whose nth member is $a(n)$ or a_n.

$$\mathrm{A}-\mathrm{B}=\{x: x\in\mathrm{A} \;\&\; x\notin\mathrm{B}\},\ \bar{\mathrm{A}}=\mathcal{N}-\mathrm{A},$$

when A is a set of natural numbers except in the proofs of lemma 16.4.1 and theorem 16.4.2. $\mathrm{A}\subseteq\mathrm{B}$ means $x\in\mathrm{A}$ implies $x\in\mathrm{B}$.

N.B. $\mathrm{A}\subset\mathrm{B}$ *always* means $\mathrm{A}\subseteq\mathrm{B}\,\&\,\mathrm{A}\neq\mathrm{B}$.

$\langle x, y\rangle$ is the ordered pair of the (natural) numbers x, y. $\mathrm{A}\times\mathrm{B}=\{\langle x, y\rangle : x\in\mathrm{A}\,\&\,y\in\mathrm{B}\}$; $\mathrm{A}^2=\mathrm{A}\times\mathrm{A}$, etc. $\cup$, $\cap$ denote union, intersection, respectively, as usual. Two sets are said to be *disjoint* if their intersection is the empty set.

$\alpha, \beta, \ldots$ denote ordinals and an ordinal is construed as its set of predecessors. With one exception, which is explicitly noted,

ordinal always means *countable* ordinal.

If A is a set of ordinals then $\sup\mathrm{A}=\lim\mathrm{A}=\cup\mathrm{A}=$ the least ordinal greater than or equal to any ordinal in A. The order type of the rationals is denoted by η as usual. η is the type of any countable dense linearly ordered set without first or last element.

$\aleph_0$ is the cardinal of $\mathcal{N}$ and $2^{\aleph_0}=c$ is the cardinal of the continuum.

A relation is a set of ordered pairs of numbers, i.e. a subset of $\mathcal{N}^2$. We use Gothic letters $\mathsf{A}, \mathsf{B}, \ldots$ for relations. By the converse, A^*, of a relation A we mean $\{\langle x, y\rangle : \langle y, x\rangle\in\mathsf{A}\}$. If A, B are relations such that $\mathsf{A}\subseteq\mathsf{B}$ then we say A is *cofinal* in the B if for some x $\langle x, y\rangle\in\mathsf{B}$ then $\langle y, z\rangle\in\mathsf{A}$ for some z. The restriction of A to B, $\mathsf{A}[\mathrm{B}=\mathsf{A}\cap\mathrm{B}^2$.

3 Functions.

A map of a (possibly proper) subset of $\mathcal{N}^n$ into $\mathcal{N}$ is called a (*partial*) *function* of n arguments. A partial function is total if it is defined on all of $\mathcal{N}^n$. We use lower case italic letters ($f, g, h, \ldots$) for functions and we write either $f(x)$ or f_x for the value of the (one-place) function f at x. The domain of a (one-place) function f is the set of numbers for which f is defined. We write δf for this set and we write ρf for the range of f, i.e. the set of values of f. By $f(\mathrm{A})$ we mean $\{f(x): x\in\mathrm{A}\}$ and similarly, if A is a relation we write

$$f(\mathsf{A})=\{\langle f(x), f(y)\rangle : \langle x, y\rangle\in\mathsf{A}\}.$$

If f is one-one then we define $f^{-1}(x)=y$ if, and only if, $f(y)$ is defined and $=x$. We sometimes talk of functions of ordinals and here we mean functions from (n-tuples of) *countable* ordinals into the countable ordinals. The classical set-theoretic definitions of addition, etc. for ordinals may be found in SIERPINSKI (1958).

4 Recursion theory.

We assume familiarity with the basic notions of partial recursive and (general) recursive functions and recursive and recursively enumerable (r.e.) sets. We sometimes use Turing machine methods for convenience (details can be found in KLEENE, 1952). The following two facts are basic:

(i) A set A is recursive if, and only if, A and $\bar{\mathrm{A}}$ are r.e.

(ii) An (infinite) set A is recursive if, and only if, it can be enumerated in (strict) order of magnitude by a recursive function.

We recall that a set containing no infinite r.e. subset is said to be *immune* and that such sets (indeed c of them) exist by DEKKER (1953). We use the well-known (primitive) recursive map $j:\mathcal{N}^2\to\mathcal{N}$ which is $1-1$ and onto, defined by

$$j(x, y) = \tfrac{1}{2}(x + y)(x + y + 1) + x$$

and also the (primitive) recursive k, l such that

$$j(k(x), l(x)) = x .$$

We write

$$j(\mathrm{A}, n) = \{j(a, n): a \in \mathrm{A}\},$$
$$j(\mathrm{A}, \mathrm{B}) = \{j(a, b): a \in \mathrm{A},\ b \in \mathrm{B}\}.$$

We use the usual (primitive) recursive function defined by

$$a * b = c ,$$

if

$$a = 2^{a_0} \cdot 3^{a_1} \cdot \ldots \cdot p_m^{a_m}, \quad b = 2^{b_0} \cdot \ldots \cdot p_n^{b_n}$$

and

$$c = 2^{a_0} \cdot \ldots \cdot p_m^{a_m} \cdot p_{m+1}^{b_0} \cdot \ldots \cdot p_{m+n}^{b_n} \quad \text{where} \quad a_m, b_n \neq 0 .$$

We make a lot of use of Kleene's indefinite description operator ν_x which from a partial recursive function f yields a partial recursive function g such that

$$(\exists y)(f(y) = 0) \Rightarrow g(\nu_y(f(y) = 0)) = 0 .$$

We define $x \dot{-} y$ as usual by

$$x \dot{-} y = x - y \quad \text{if} \quad x \geq y,$$
$$= 0 \qquad \text{otherwise},$$

but we also define

$$x \div y = \mu_z \{y + z = x\}.$$

So $x \div y$ is a partial function in general.

Other unexplained notations may be found in KLEENE (1952) especially p. 538.

REFERENCES

P. H. G. ACZEL, 1966, D. Phil. Thesis, Oxford.

1966a, *Paths in Kleene's O*, Archiv Math. Logik Grundlagenforschung **10**, 8–12.

— & J. N. CROSSLEY, 1966, *Constructive Order Types*, III, Archiv Math. Logik Grundlagenforschung **9**, 112–116.

H. BACHMANN, 1955, *Transfinite Zahlen* (Berlin).

G. CANTOR, 1915, *Contributions to the Founding of the Theory of Transfinite Numbers*, (Dover Reprint).

A. CHURCH & S. C. KLEENE, 1936, *Formal Definitions in the Theory of Ordinal Numbers*, Fundamenta Mathematicae **28**, 11–21.

J. N. CROSSLEY, 1963, D. Phil. Thesis, Oxford.

—, 1965, *Constructive Order Types*, I, in *Formal Systems and Recursive Functions*, Eds J. N. Crossley & M. A. E. Dummett (Amsterdam) 189–264.

—, 1966, *Constructive Order Types*, II, JSL **31**, 525–538.

— & R. J. PARIKH, 1963, *On Isomorphisms of Recursive Well-orderings* (Abstract), JSL **28**, 308[171].

— & K. SCHÜTTE, 1966, *Non-uniqueness at ω^2 in Kleene's O*, Archiv Math. Logik Grundlagenforschung **9**, 95–101.

J. C. E. DEKKER & J. MYHILL, 1960, *Recursive Equivalence Types*, Un. of California Publications in Mathematics, n.s. **3**, 67–214.

A. EHRENFEUCHT, 1957, *Applications of games to some problems of mathematical logic*, Bull. Acad. Polon. Sci. **5** pp. 35–37.

E. ELLENTUCK, 1963, *Solution of a problem of R. Friedberg*, Mathematische Zeitschrift, **82**, 101–103.

S. FEFERMAN, 1968, *Systems of Predicative Analysis*, II; JSL **33**, 193–220.

R. FRIEDBERG, 1961, *The Uniqueness of Finite Division for Recursive Equivalence Types*, Math. Zeitschrift **75**, 3–7.

R. O. GANDY, 1960, *Proof of Mostowski's Conjecture*, Bull. Polon. Acad. Sci. **8**, 571–575.

A. G. HAMILTON, 1968, *An unsolved problem in the theory of constructive order types*, JSL. **33**, 565–567

S. C. KLEENE, 1952, *Introduction to Metamathematics* (Amsterdam).

—, 1955, *On the Forms of Predicates in the Theory of Constructive Ordinals* (Second Paper), Am. J. Math. **77**, 405–428.

G. Kreisel, 1960, *Non-uniqueness Results for Transfinite Progressions*, Bull. Polon. Acad. Sci. **8**, 287–290.

J. McCarthy, 1956, *The Inversion of Functions defined by Turing Machines*, Automata Studies, Annals of Mathematics Study no. 34, (Princeton) 177–181.

A. Nerode, 1961, *Extensions to Isols*, Annals of Mathematics **73**, 362–403.

R. J. Parikh, 1962, *Some Generalisations of the Notion of Well-ordering* (Abstract), Notices Am. Math. Soc. **9**, 412.

—, 1966, *Some Generalisations of the Notion of Well-ordering*, Zeitschrift Math. Logik Grundlagen der Mathematik **12**, 333–340.

H. G. Rice, 1956, *Recursive and recursively enumerable orders*, Transactions Am. Math. Soc. **83**, 277–300.

K. Schütte, 1965, *Predicative Well-Orderings*, in *Formal Systems and Recursive Functions*. Eds J. N. Crossley and M. A. E. Dummett (Amsterdam), 280–303.

W. Sierpinski, 1948, *Sur la Division des Types Ordinaux*, Fundamenta Mathematicae, **35**, 1–12.

—, 1958, *Cardinal and Ordinal Numbers*, (Warsaw).

R. I. Soare, 1969, *Constructive order types on cuts* (to appear in JSL).

C. Spector, 1955, *Recursive Well-Orderings*, JSL **20**, 151–163.

A. Tarski, 1949, *Cardinal Algebras*, (New York).

—, 1956, *Ordinal Algebras*, (Amsterdam).

J. S. Ullian, 1960, *Splinters of Recursive Functions*, JSL **25**, 33–38.

O. Veblen, 1908, *Continuous Increasing Functions of Finite and Transfinite Ordinals*, Transactions Am. Math. Soc. **9**, 280–292.

A. N. Whitehead & B. Russell, 1927, *Principia Mathematica*, Vol. II (Cambridge).

NOTES

INTRODUCTION

01 This is closely related to the results of CROSSLEY AND PARIKH (1963).
02 The term "recursive isotonism" which has been used instead of "recursive isomorphism" in CROSSLEY (1963, 1965), ACZEL AND CROSSLEY (1966) and some other places has a slightly different meaning.
03 This will be exploited more in work by NERODE and the author which is in preparation.
04 We abbreviate "recursively enumerable" by "r.e.".

CHAPTER 1

11 The author has discovered (July 1966) that this theorem was essentially proved by RICE (1956, theorem 20) and an essentially identical one by SPECTOR (1955).
12 There will be no confusion between sequence numbers of sequences of two elements and ordered pairs since the context will always make clear which is intended. $*$ is defined on p. 213.

CHAPTER 2

21 Note that this notation will cause no confusion for if A is set-theoretically included in B then $A = B$.
22 We use the word "step" here in the sense of a whole phase in the calculation rather than moving just one square on the Turing machine tape.

CHAPTER 3

31 The quantifier $(\forall f)$ ranges over all one-place functions and the quantifier $(\forall B)$ ranges over all sets of natural numbers.
32 This use of the word "splinter" is derived but differs from that in ULLIAN (1960).

CHAPTER 4

41 Such exists by theorem 3.1.5.

CHAPTER 5

51 Although these are not quite the same orderings as in CROSSLEY (1965), the co-ordinals are the same.
52 Since we get most of our counterexamples from V.
53 This simple group-theoretic way of presenting the proof is due to Alex Rosenberg.

CHAPTER 6

61 By "minimal" we mean minimal with respect to domain and range.
62 This argument is basically due to TARSKI (1956).
63 This argument and those to the end of this chapter are due to TARSKI (1956) though the proofs of the supporting theorems are very different from his.

CHAPTER 7

71 Since we are assuming min $\mathsf{A} = 0$ we shall never misidentify two sequences since no sequence can end in 0.
72 This is a strengthened version of theorem VIII.2.2 of CROSSLEY (1965).

CHAPTER 8

81 This is inspired by TARSKI (1956). Compare also theorem 2.4.11.
82 We are here using an extension procedure similar to that in the proof of theorem 8.3.5.
83 We write $e(\mathsf{A}, \mathsf{B})$ for $\{e(A) : A \in (\mathsf{A}, \mathsf{B})\}$.
84 Recall $(x)_0 =$ exponent of 2 in prime factorization of x.

CHAPTER 9

91 In fact we show that many co-ordinals may be expressed as polynomials in W with large exponents.
92 This theorem was first conjectured by A. L. Tritter.
93 We are assuming $\min(\mathsf{A}) = 0$ as usual.

CHAPTER 10

101 This definition is adapted from KREISEL (1960).
102 The problem of whether there exist principal numbers for multiplication not of the form W^A is solved affirmatively in the appendix B by A. Hamilton.

CHAPTER 11

111 The results in this chapter were obtained by P. H. G. Aczel and the author and

appeared in their original form in ACZEL and CROSSLEY (1966). The name "E-number" is intended simply to convey that these co-ordinals are closely related to (classical) ε-numbers. It should not be confused with the E(1), etc. of VEBLEN (1908).

112 $\varepsilon_\alpha = \mu_\beta\{\omega^\beta = \beta \& \beta > \varepsilon_\gamma \text{ for } \gamma < \alpha\}$ or, equivalently, the α-th (classical) principal number for exponentiation greater than ω.

113 Here $x_1 \geq_{\boldsymbol{E}} \ldots \geq_{\boldsymbol{E}} x_r$ is an abbreviation for

$$\langle x_2, x_1\rangle \in \boldsymbol{E} \& \ldots \& \langle x_r, x_{r-1}\rangle \in \boldsymbol{E}.$$

CHAPTER 12

121 $\bar{y}$ is defined on p. 14.

CHAPTER 13

131 The reader who is familiar with R.E.T.s and isols is advised to omit this chapter.

132 There will be no confusion although we use the same type founts for (e.g.) R.E.T.s and C.O.T.s since the context will make clear which is intended (generally, R.E.T.s in this chapter).

133 Here again we are dealing with *unordered* sets so $\simeq$ is not ambiguous.

CHAPTER 14

141 The construction given here is due to C. G. Jockusch.

APPENDIX

A1 The question whether the theorem holds with the *weak* sense of $\leq$ is open.

REFERENCES

171 See the correction in the author's abstract JSL **31**, 292–3.

INDEX OF SYMBOLS

INDEX OF TERMS